A DAVID & CHARLES BOOK
Copyright © David & Charles Limited 2011

David & Charles is an F+W Media Inc. company
4700 East Galbraith Road
Cincinnati, OH 45236

First published in the UK in 2011

Text copyright © F+W Media Inc. 2011

The material in this book has been previously published in *Why Didn't I Think of That?* published by Adams Media, 2010.

F+W Media Inc. has asserted his right to be identified as author of this work in accordance with the Copyright, Designs and Patents Act, 1988.

All rights reserved. No part of this publication may be reproduced, stored in a retrieval system, or transmitted, in any form or by any means, electronic or mechanical, by photocopying, recording or otherwise, without prior permission in writing from the publisher.

A catalogue record for this book is available from the British Library.

ISBN-13: 978-1-4463-0043-5 paperback
ISBN-10: 1-4463-0043-9 paperback

Printed in China by RR Donnelley
for David & Charles
Brunel House, Newton Abbot, Devon

Senior Acquisitions Editor: Freya Dangerfield
Desk Editor: Felicity Barr
Project Editor: Stuart Robertson
Proofreader: John Skermer
Design Manager: Sarah Clark
Production Controller: Bev Richardson

David & Charles publish high quality books on a wide range of subjects. For more great book ideas visit: www.rubooks.co.uk

Contents

Introduction	004
ONE: THE BIG ONES	006
TWO: STUFF YOUR FACE	016
THREE: FUN AND GAMES	034
FOUR: HOME AND OFFICE ESSENTIALS	056
FIVE: ACCESSORIZE THIS!	092
SIX: TOOLS AND TECHNOLOGY	108
SEVEN: A MATTER OF LIFE AND DEATH	136
Resources	154
Appendix	155
About the author	156
Index	157

Introduction

THE LIGHT BULB
THE CAR
THE THEORY OF RELATIVITY
THE STEAM ENGINE
THE COMPUTER

What do all of these revolutionary inventions have in common? None of them are in this book. Why? Because the average person can't invent stuff like that and this is an 'invention book' for the average person.

I have compiled the stories behind the most extraordinarily, yet simple inventions that have changed our world. In fact, in most cases, the astounding inventions described here required no special skills, no education, no expensive laboratories, no government grants, very little capital, just an extraordinary amount of hard work and ingenuity. This book is meant to inspire you, make you laugh a little (I hope) and encourage you through example to follow your dreams.

However, just because they're identified as simple, doesn't mean the inventors are simple-minded. Every one of these inventions is elegant – meaning, they all have refinement, grace and beauty. The people who invented and discovered these things deserve to be recognized and praised.

All that said, you will see that there are inventions and discoveries listed here that would be impossible for you to have created, as their inceptions occurred long before you were born. Fire, the wheel, even golf – these are all amazingly simple but now out of your creative reach. Inventions and discoveries such as these are rendered here to display how, throughout history, ordinary people did extraordinary things time and time again.

But you will also find an astonishing array of modern items that you really could have invented had you been the first to think of them. So turn the page, read on and as you do, keep asking yourself, 'Bollocks! Why didn't I think of that?'

THE BIG ONES

Fire

TAGLINE: fire
PREDECESSOR: cold
LESSON: cold bad, fire goooood

While fire may technically be more of a discovery than an invention, this book would be incomplete without its inclusion. It is thought that the discovery of fire most likely happened when a lightning strike ignited a tree or some brush during prehistoric times, though it's impossible to say exactly how it came about. I'd like to think it went a little something like this...

'HUMAN SUBTLETY ...

will never devise an invention more beautiful, more simple or more direct than does nature, because in her inventions nothing is lacking, and nothing is superfluous.'

Grog: Light from sky make tree burn!
Steve: Fire burn squirrel in tree.
Grog: Me eat squirrel. Pass salt.
Steve: Salt not been discovered yet.
Grog: Grrr!

DID YOU KNOW?

Heat + oxygen + fuel; these are the three ingredients to a fire, as identified by fire-fighting professionals. Remove one of these ingredients and out goes the fire.

It's believed that the controlled use of fire started during the early Stone Age. Some of the earliest evidence of Grog and pals getting their barbecue on was discovered at an archaeological dig site in Benot Ya'aqov, Israel, where charred seeds and a YOU KISS COOK NOW! apron* were uncovered, dating back 790,000 years.

Controlled fire is arguably man's most significant discovery, pertinent for warmth, illumination, and, of course, adequately describing the pants of a liar-liar.

*I made that up.

Wheel

TAGLINE: a round cylinder

PREDECESSOR: dragging stuff

LESSON: you may be looking at the greatest invention ever without realizing it

I get it – the wheel is probably the most important mechanical invention of all time. But come on! How hard was it to suss out? I mean even during the Stone Age stuff rolled, didn't it? Some caveman must've noticed a log or a rock rolling down a hill. It probably happened all the time! However, it took quite some time to work out the principle that round things roll. And for many cultures it eluded them entirely until nearly modern times! Because the concept behind the wheel appears to be so simple, one might assume that people would naturally pick up on it. But incredibly, this is not the case. The Inca, Aztec and Maya cultures were highly advanced, yet they never used the wheel. In fact, there is no evidence that the wheel existed in American indigenous civilizations until after the Europeans came over to 'befriend' them.

The very first wheel was most likely made of a section of wooden log. Later, more advanced tools made it possible to carve wheels from stone. The first flat tyre followed shortly thereafter (only kidding). Today the wheel is so commonplace it is barely noticed. However, aside from the obvious applications for transport vehicles such as cars, trucks and trains, etc., we are dependent upon the wheel in many less obvious ways.

In fact, nearly every piece of machinery created uses a form of wheel in its mechanisms. From watches to turbine engines, modern society would grind to a halt like a rusty gear without this simplest of contraptions.

Did you know?

The classic spoked and hubbed wooden wheel with an iron rim was in widespread use for nearly three millennia – from Iron Age Europe through to the late twentieth century, and is still found in many rural parts of the world.

Spear

TAGLINE: a pointy stick
PREDECESSOR: a stick with less poking ability
LESSON: monkeys are like really hairy people

Here's an invention so simple that even monkeys use it. Think hunting and warfare are just human pastimes? Think again.

The common chimpanzee has been known to manufacture and use the spear. Chimpanzees near Kédougou, Senegal, have been seen fashioning spears by breaking limbs from trees, sharpening one end with their teeth and then, apparently, using the weapons to hunt galagos, a lemur-like primate. Orang-utans have also been witnessed using spears. Rather than hunt other primates though, they use the tool to fish, having observed humans fishing in similar ways. Clever buggers!

It's been suggested that the discovery of the primates' use of spears means early humans probably used spears as well, perhaps even as early as five million years ago. And they have been used for hunting, defence and warfare ever since. It's just a pointy stick, but somebody had to think of it. And if a monkey can invent stuff, cheer up, so can you! It won't be long 'til they're picking up tennis racquets, you know. And then we'll be obliged to let Alan Partridge know.

> **Did you know?**
>
> The spear is a symbol of power in Chinese martial arts. It's referred to as the king of weapons.

Knife

TAGLINE: sharp object that cuts, stabs and butters

PREDECESSOR: ripping stuff

LESSON: you have to be sharp to invent sharp things

The knife! It slices! It dices! It can cut through a tin can and still make a clean cut through a tomato! But wait! There's more... Knives have always been valuable tools. They started out being made from chipped volcanic rock as far as 2.6 million years ago, with artefacts being found in the Afar region of Ethiopia. Similar tools were made throughout the Palaeolithic era from animal bone and wood. And up until relatively recently – about ten thousand years ago – these were the main materials used to make knives.

Incredible, but true! Still not convinced that the knife is one of the most useful tools ever invented? Wait 'til you hear this! The knife improved with advances in metallurgy; knives made of bone and wood became a thing of the past. Now it was all about the cutting power of steel.

Cold, hard steel. And ever since then they've been the go-to item no matter what the need – whether it's eviscerating your mortal enemy or buttering your toast.

They slice! They dice! They stab! They poke! They slather! Knives! Now how much would you pay? Wait! Don't answer that! There's still more! Today we slice, dice and scrape with bread knives, diving knives, hunting knives, sabres, cutlasses, livestockman knives, scalpels, utility knives, cooking knives, filet knives, Bowie knives, carving knives, bread knives, pocket knives, flick knives, fish knives, electric knives... the list goes on! Available at a certified dealer near you! Act now!

That's not a knife – *that's* a knife

When does a big knife cease to be a knife and become a sword? Well, it's not simply a matter of size. Function is the key here. A sword is primarily designed to slash; a knife is designed to stab or puncture. In medieval Germany, common peasants were restricted from carrying swords. So they just made their 'knives' bigger – up to two or three feet in length.

Gravity

TAGLINE: a miniature nail

PREDECESSOR: the nail

LESSON: make a 'light' version of something that's already useful.

Here's a discovery that's not terribly complicated. It goes like this: Sir Isaac Newton said we stick to the ground. Einstein said we are pushed to the ground.

Either way, we don't sail into space. And it's all thanks to gravity. While you might be quick to think Newton's the 'Gravity Guy', you'd be mistaken. The theory behind this principle started to formulate in the mind of Greek philosopher Claudius Ptolemy in the second century. The next notable stab at it was by Nicolas

I can see Uranus

Newton's theory made its biggest splash when it helped to prove the existence of Neptune based on the motions of Uranus.

> **Did you know?**
>
> In 1952 Albert Einstein was offered the presidency of the newly formed state of Israel. He declined.

Copernicus in the early 1500s. By the late 1500s the theory (not yet called gravity) really began to gain traction when people started to notice Galileo Galilei's balls. Galileo was showing everybody his balls. And why not? They were magnificent and he was doing exceptional things with them. Famously, the Italian scientist started his theorizing when he dropped his balls from the Tower of Pisa.

His study of falling bodies and objects rolling down inclines paved the way for Newton and his apples. It is rumoured that Newton discovered gravity when an apple fell from a tree and landed on his head. Whereupon William Tell shot it with a crossbow. Is that right? Probably not. Anyway – while there are actual accounts of Newton mentioning a falling apple when he was formulating his theory, the fruit did not actually strike the scientist. He merely used it as an example when describing his findings.

STUFF YOUR FACE

Sliced Bread

TAGLINE: the greatest thing since itself

PREDECESSOR: slicing your own bread

LESSON: improve something that people already want, and, never give up

Sometimes the best inventions are ways to slice other inventions into smaller pieces. Wait... that didn't come out right. What I mean is, sometimes the best inventions are modifications of other inventions. That's better. In this case, we're talking about modifying an invention that has, literally, fed the world for centuries – bread.

Former jeweller Otto Frederick Rohwedder came up with the idea in 1917. And it only took him seventeen years to sell it. In addition to some personal and business problems, Otto had to overcome the doubts of bread-makers who were uncertain that people would want to buy pre-sliced bread. They rightfully surmised that the loaves would get stale quicker that way. Otto solved this problem by packaging the bread immediately after slicing it. In 1928, sliced bread began its rapid climb toward being the norm.

So why has sliced bread become the standard to which all other inventions are held? Is it because it added a level of convenience to a product that was such a staple in the world's diet? Is it because it arrived at a time in our history when timesaving was at a premium? Nope.

In 1930, American brand Wonder Bread coined the saying 'The greatest thing since sliced bread' in an advertising campaign. The slogan stuck and has become a way to applaud an invention. Makes you wonder how they described Otto's first loaf.

Mmmm, protein as well as carbohydrate and fibre!

In September 2010 Britain's sandwich-loving community was shaken to its core when a man wanting a lunchtime snack found what he thought was a lump of uncooked dough in the crust of his loaf. Making several rounds of sarnies, he eventually worked his way down to the end, whereupon to his delight he found a partially sliced baked mouse embedded into the bread. The hungry man was absolutely livid. The mouse's opinion wasn't canvassed, but he looked pretty relaxed about the whole thing.

Ice Lolly

TAGLINE: a sweet frozen treat
PREDECESSOR: juice
LESSON: freeze treats

How many times have you placed a lukewarm drink in the freezer to cool it down only to forget about it and find it frozen solid hours later? An innocent and annoying mistake to you, a gold mine to one eleven-year-old Frank Epperson. In 1905, he left his drink outside overnight on the porch of his California home. It froze and the stick he had used to stir it stuck in place. So did the idea. He named it the Epperson icicle, which of course didn't stick.

The following summer, he recreated the treats in his family's fridge and sold them around town, under the shortened name 'Epsicle'. Incredible it seems to us today, but Epperson finally patented his treat in 1924 under the name 'Popsicle', which came from his children's frequent requests for their Pop's sickles.

> **Did you know?**
>
> By 1928, Epperson had earned royalties on more than 60,000,000 ice lollies. But think of the number of piss-poor jokes he would've had to come up with for the lolly sticks.

By 1928, Epperson had sold the rights to the name Popsicle and earned royalties on more than sixty million ice lollies. That's a pretty cool profit for accidentally leaving a drink on your back porch. Makes you wonder how much money some of your stupid mistakes could make you, doesn't it? Today, lollies are made and distributed by a number of different companies in a number of different shapes and flavours. Happily, buried under all the expensive luxury choc 'n' caramel meals-in-one clogging up the newsagent's freezer you can still find good old-fashioned frozen E-numbers lurking about.

Turducken

TAGLINE: an ungodly combination of birds
PREDECESSOR: eating poultry one species at a time
LESSON: no idea is too weird

A bloke walks into a butcher's shop with a turkey, a chicken and a duck... No, this is not a joke – it's the turducken.

Not since the 'refried bean' has boredom been the catalyst for such culinary delight. Three boned, stuffed birds, crammed into one another (I've seen videos on the internet like that): a turkey stuffed with a duck, stuffed with a chicken. The turducken is food gone a-fowl.

A meal consisting of three life forms, fused, as if by some hideous matter transfer experiment involving Jeff Goldblum, may seem more like a zoological discovery than an invention, but I assure you, the turducken does not occur naturally in the wild.

The concept of shoving dead meat inside bigger bits of dead meat has been around for years. British and European gastronomes/greedy bastards have been ramming between five and seventeen different types of birds inside each other and then baking or roasting the lot for at least three centuries, usually to mark a special event. But what of your standard-issue turducken itself? Apparently, one day in 1985, a man whose name has been lost to history strolled into Hebert's Specialty Meats in Maurice, Louisiana, carrying three birds and instructed the staff there to build him one turducken. Mercifully, the procedure was performed post-mortem. Once word got out that there was a way to kill, mangle and then eat three animals at once, the demand went through the roof, particularly round Thanksgiving in the US.

If ever you blew a chance to invent something when you were high and hungry, this was it. But, hey, it's not too late. Surely you can cram a Cornish game hen into that chicken and then maybe a parakeet into that? Good luck.

Did you know?

The turducken does not occur naturally in the wild. Neither does the haggis, though that doesn't prevent one-third of US visitors to Scotland believing otherwise, according to a 2003 report.

French Fries

TAGLINE: potato slices thrown into boiling oil
PREDECESSOR: whole potatoes thrown into boiling water
LESSON: if you deep-fry it, they will come

The French insist that they invented the French fry – the Belgians say it was them. So in the interest of fairness, I called the Belgian Embassy and enquired about its creation.

Based on my extensive research, which consisted of tricking a Belgian receptionist into recognizing fried potato strips as French fries, I'm going to go with the French side of the story. Plus, the Belgians already have that delicious waffle named after them.

And on top of my conversation with Océane, dead US president Thomas Jefferson also backs up the French claim. Now Jefferson wasn't 'Washington honest' – as evidenced by all those slave children who looked like him – but there isn't any evidence that he lied about fried foods. A menu was found from one of Jefferson's dinner parties held in 1801, which clearly states that his guests would be served 'potatoes, deep fried and served in the French way'. This was risky because potatoes back then were thought to be highly poisonous unless boiled thoroughly. Jefferson assured his guests that his French chef, Honoré Julien, would prepare the potatoes in a manner that would not kill them and they did not.

Yes, it seems it was a Frenchman who came up with the extraordinarily simple revelation that potatoes taste good after being fried in a vat of oil.

But then, doesn't everything?

Belgian Embassy: Hello, Belgian Embassy, this is Océane. How can we help you?

Me: Quick question, Océane. Who invented the sliced potato boiled in oil?

Océane (in a cute accent): Ah, excellent question. Belgium has historical evidence that we were eating potato strips fried in oil as early as the seventeenth century.

Me: You mean, French fries?

Océane: Oui.

Me: Ah ha!

Chocolate Bar

TAGLINE: chocolate in the shape of a bar

PREDECESSOR: smaller pieces of chocolate

LESSON: sometimes a lot of people can make a lot of money from one idea

Before being messily slaughtered by Spanish invaders, Aztec Emperor Montezuma loved sipping on his favourite chocolate drink made from cocoa. Conquistador Hernando Cortez, the man behind Montezuma's killing, brought the drink to Spain in 1529. It became an instant favourite of the Spanish royalty and later spread to the rest of Europe where it became equally popular. It took three centuries on the continent before it was turned into a solid confection.

In 1847, the English company Joseph Fry & Son discovered a way to mix a chocolate-drink-based concoction into a paste that could be pressed into a mould. Et voilà! The chocolate bar

Did you know?

The Snickers bar is the bestselling chocolate bar in the world, with annual global sales of £1.25 billion. The memory of the poor old Marathon lives long, though...

was born! Soon people began eating chocolate as much as drinking it. Brummie John Cadbury began selling a similar version of the Joseph Fry & Son chocolate bar in 1849. However, the early bars of chocolate were bittersweet. To solve this problem, Swiss bloke Henri Nestlé and Daniel Peter introduced milk chocolate in 1875.

Today, a billion kilos of chocolate are produced each year. And believe it or not, the good ol' guzzlers of the United States don't take top honour in its consumption. According to WeAreSweet.com, the Swiss consume the most chocolate. Well, you've got to have something to eat while you're busy hoarding money and admiring over-engineered clocks.

Plastic Milk Crate

TAGLINE: furniture for students and picket-liners

PREDECESSOR: wooden crates

LESSON: useful objects may have other uses – think ahead and capitalize on them

The story of the milk crate isn't so much one of invention but of reinvention.

As is often the case with a good idea, the milk crate was created to do one thing but ended up becoming popular for quite another.

A pint of lo-fi reggae please

Milk crates have long been in demand as low-budget, portable storage for 12" vinyl records. According to the word on the street round the back of Sainsbury's delivery bay, the ones from the seventies and eighties are the best fit. Well, that's what people stuck in the seventies and eighties always say, isn't it? Now pass me my UB40 best of.

Obviously, the milk crate was invented to transport milk. And originally, it was crafted out of wood or metal. However, the 1960s brought about the plastic crate and its rise in non-dairy-related uses. From functioning as shelves to bed-raisers to stools, milk crates became the 'in' home furnishing for the young-and-skint. Its new use as the perfect piece of flexible furniture caused not only a rise in its popularity but also in the rate at which it was pinched.

Until recently, the majority of the pilfering was blamed on students who use milk crates to construct just about every piece of furniture one can imagine. Recently though, it has been discovered that milk crates are being stolen by actual thieves, ground up, sent to China and used in the creation of a wide variety of goods. Yeah, apparently there's a milk crate mafia! Because of this, dairies now often hire private investigators to root out the milk crate marauders. Think that's a bit on the excessive side? Milk crate thefts cost dairies millions every year.

So the next time you try to nick a milk crate from round the back of the Co-Op, remember: somebody's got to pay for it. And if you're snagged by an in-store private dick, it might just be you. Could you stand the shame?

Bottled Water

TAGLINE: free water that you buy

PREDECESSOR: free water

LESSON: there's no such thing as a stupid idea and *sell free stuff*

It might be easier (and funnier) to dismiss bottled water as a brilliant marketing ploy, but there was actually a real need that fuelled the industry. Seems some people didn't like dying gruesome deaths from cholera, E. coli and other diseases. Pansies.

Even as early as the dawn of the Roman Empire, spring water was being routed into the city of Rome from miles away by aqueducts. A few centuries later, bottled water became a serious moneymaking industry once glass bottling made the process more cost-efficient.

In the nineteenth century it became a booming industry, particularly in France around the Alps and Auvergne where famous brands like Evian and

> By 1856, US company Saratoga Springs was producing nearly 7 million bottles per year.

Volvic benefited from official recognition of mineral water's healthy properties through the establishment of the Société des Eaux Minerales.

The boom subsided a bit when municipalities started adding chlorine to public drinking water (mmm! chlorine), taking away the health risk associated with filling your glass at the tap. However, there was a second bottled water boom in 1977 thanks to Perrier's yuppie-targeted marketing. The multi-million pound campaign made bottled water as essential in yuppie culture as the BMW. And as Dr Francis Chapelle notes in his book *Wellsprings: A Natural History of Bottled Spring Waters*, Perrier 'was all the things yuppies wanted in a lifestyle-defining product.'

These days, Americans consume the most bottled water in the world, despite the fact that nearly every person in the United States has access to perfectly safe drinking water.

Straw

TAGLINE: furniture for students and picket-liners
PREDECESSOR: wooden crates
LESSON: useful objects may have other uses – think ahead and capitalize on them

The earliest drinking straws were made from hollowed out pieces of grass and reed. Worked out how the invention got its name yet? In 1888, Marvin Stone sought to improve on the natural straw by patenting the paper straw. However, this 'improvement' was actually inferior.

Why? It was made of an absorbent material! A little detrimental if you're hoping to have liquid travel through it. Every sip a person took damaged the straw. Inefficient and wasteful – so it was back to the drawing board.

Did you know?

Using a straw when drinking a sugary drink helps reduce tooth decay. Some people claim that if the drink also contains alcohol, drinking through a straw may increase the rate at which you get drunk. But only because you drink quicker.

It was a combination of Stone's patent and non-absorbent plastic that got us to the effective device we use today. But the brainstorming didn't stop there. Now that the straw could actually be used, inventors had a field day with the different types of straws that could be created. Stone's fallible paper straw led to:

- The bendable straw, with its flexible neck
- The crazy straw, with tons of twists and turns for your drink to take, popular in the 1980s
- The spoon straw, perfect for scooping the last bit of slush or milkshake
- The mini-straw, which comes attached to your favourite carton drink
- The edible straw, made out of food like chocolate and cereal

There were a few dips in the process, but the straw has come a long way from the piece of grass it once was.

Shopping Trolley

TAGLINE: a basket on wheels

PREDECESSOR: er... a normal basket

LESSON: make it easier for people to buy stuff and they'll buy more stuff

The long, wobbly-wheeled road started in 1936 when American grocery store owner Sylvan Goldman put the first shopping carts into service. Goldman was trying to work out how to make the shopping experience more efficient so customers would spend more money. So he took a folding chair, put a basket on the seat, wheels on the legs and – wham! – the shopping trolley was born.

Do fish go shopping?

According to a 2009 national survey of British waterways, over 3000 shopping trolleys were sent for an early bath in choked urban canals that year. The cost of recovering them was estimated at £150,000. Obviously it's a dumbass thing to do, not least because it clogs up vital havens for wildlife and recreation. But don't weep too much for the supermarkets – the cost equates to just 18 minutes' worth of annual profits.

Simple enough, right? Wrong.

It seems shoppers back then didn't take to the 'carts' as well as we do now. Blokes thought it was rather sissy to push around a trolley while young women found it unfashionable. And old people thought it made them look feeble, which they probably were so that was a pointless objection. (Makes you wonder what they'd think about the motorized scooter version you see today.) But Goldman wasn't about to give up. Like many of the inventions included in this book, marketing played a big part in the shopping trolley's popularity.

Goldman hired models – trolley-dollies if you will – to make pushing shopping carts look cool. And it worked. By 1940, shopping trolleys were so common that supermarkets were redesigned to accommodate them. Nowadays you can't go into a shop without seeing them clogging up the aisles, or picking that one with the squeaky, demented wheel. Apparently they only cost a quid though, so what do you expect?

FUN AND GAMES

Yo-Yo

TAGLINE: a toy you toss and it returns
PREDECESSOR: a toy you toss and it breaks
LESSON: deadly weapons do make great kids' toys

Allow me to apologize in advance for the following: the yo-yo has had its ups and downs.

Once again we find ourselves examining ingenuity through the recognition of a good idea, rather than the actual invention of an object. A lot of people assume that some American bloke invented the yo-yo in the 1950s. But he most certainly did not. In fact, the yo-yo had been around for over two centuries before it became popular in the West.

It first made its way to the Western world in the 1800s, where the British referred to it as the quiz and bandalore, which was adopted from the French who also called the ancient toy an incroyable. Clearly not as catchy as yo-yo. So who came up with the name we all know it by?

> Approximately 45 million yo-yos were sold in 1962. Top five yo-yo tricks that soon evolved included 'walking the dog', 'the flying saucer', 'around the world', 'rock the baby' and 'chop sticks'.

It's actually the term used in the toy's native Philippines and translates to 'Come back!' Even better, the toy was actually used as a weapon on the island nation for over 400 years. A little different to the one you use to 'walk the dog', their version came complete with sharp edges and points – perfect for flinging at enemies and live stuff they wanted to eat. Granted, it didn't stand a chance against the colonizer's weapons, which is probably why it's no coincidence they were a conquered people for about 400 years.

But back to the yo-yo you know today... it took a man called D F Duncan's recognition of the toy's potential in order for it to become a must-have for every American kid. The company's advertising campaigns caused millions to flock to stores in search of the deadly-weapon-turned-child's-toy. The yo-yo's popularity peaked in 1962 when forty-five million of them were sold, though it remains a favourite for the four-foot set. Unfortunately, mismanagement bankrupted Duncan's company around this same time. So Duncan cut the string (and his losses). The Flambeau Plastic Company dropped in and bought Duncan's shares as well as the rights to his name and his trademarks in 1968.

Hula Hoop

TAGLINE: an excuse to gyrate...well another excuse to gyrate

PREDECESSOR: hula-ing without the hoop

LESSON: you don't have to invent it to make millions from it

Think the Hula Hoop was a '50s thing? Guess again. Only the marketing ploy came from that decade. In fact, people have been gyrating with circular hoops for some time now. The idea can be traced all the way back to ancient Egypt where children played with large rings of dried grapevines. And the first hoop trend didn't hit during the doo-wop era; it actually happened in fourteenth-century England. Doctors' records from that period credit the craze as the cause of back pains and heart attacks.

Even the term 'hula hoop' wasn't a '50s original. That happened in the 1800s when British sailors were the first to attach 'hula' to 'hoop'. After visiting Hawaii, they noted the similarity between the natives' hula dancing and their own gyrations. The natives subsequently slaughtered the sailors, but that's a whole other

> Wham-O sold 25,000,000 hula hoops in the first four months of sale.

story... The real innovation in this case isn't the actual invention, but the recognition that the concept had sales potential. And this recognition came from where else but Wham-O – a California toy manufacturer that appears repeatedly throughout this book. Wham-O marketed the hell out of the Hoop; their salesmen literally hit the pavement as they began holding Hula Hoop demonstrations at playgrounds throughout the country. Shortly after Wham-O's campaign began in 1958, the concept caught on and spread – fast. Twenty-five million Hula Hoops flew off the shelves in the first four months alone.

Like any fad, though, the Hula Hoop craze cooled. By the early '60s kids were clamouring for the next big thing. However, that doesn't mean the Hoop didn't stick around. Even today, they're still being used in back gardens and playgrounds. Think about it: Who hasn't Hula Hooped? My money's on the Archbishop of Canterbury but who can tell for sure?

Smiley Face Icon

TAGLINE: a circle with two dots and a semicircle

PREDECESSOR: fake cheerfulness without an easily identifiable symbol

LESSON: be in the business of emotion, but don't be emotional about business

In 1963, Richard Ball, co-owner of an advertising and PR firm in Massachusetts, conceived the single most used and recognizable graphic icon of the past century – the smiley face.

Says its inventor: 'I made a circle with a smile for a mouth on yellow paper, because it was sunshiny and bright. Turning the drawing upside down, the smile became a frown. Deciding that wouldn't do, I added two eyes.' It took ten minutes.

Yep. That's it.

Did Ball do it to spread a message of peace? To promote tolerance and understanding? To spread joy among his fellow man? No.

Beat Dis

In the UK the smiley icon took on a renewed significance during the 'Second Summer of Love' in 1988-89, when it came to symbolize the recreational use of MDMA ('Ecstasy') at acid house and rave venues throughout the country.

The icon was commissioned and paid for by an insurance company to help ease the bickering among its staff in the wake of a company merger. The insurance company ordered a 'friendship campaign' and hired Ball to design an image for the buttons being passed out. And that was only the beginning.

In September 1970, the smiley face left the corporate world and hit the streets. Brothers Murray and Bernard Spain of Philadelphia set out to capitalize on the growing 'hippie market' so they 'borrowed' Ball's design. The two began manufacturing the smiley button for general consumption and, boy, did it take off. Within two years, more than one hundred million buttons had been sold. And that simple smile continues to grace products and blonde receptionists' dotted 'i's to this day.

So, yes, the graphic that we use to symbolize peace, joy, happiness and bliss was the love child of two merging insurance companies that hated each other... until it was stolen and used to exploit hippies.

Have a nice day! ☺

Frisbee

TAGLINE: an upside-down plastic plate

PREDECESSOR: projectiles with no whimsical floating abilities

LESSON: find a new use for an old object

It's 1947... a Yale University student is looking for reasons to avoid studying. Suddenly, flying toward him in a graceful arc, like nothing he's seen before, is a cylindrical disc sailing effortlessly through the air. He picks it up and throws it back and he's found the perfect excuse to procrastinate.

It all started with a baker named William Russell Frisbie who had been selling his pies to students at universities and schools in New England for some time. Unsurprisingly, the students loved the pies and therefore had a lot of empty pie tins laying about with the words 'FRISBIE'S PIES' emblazoned on them. Students are a resourceful bunch: these guys started using the 'Frisbie' pie pans in much the same way you see frisbees used today. And it caught on, big-time! In 1946, two partners, Walter Frederick Morrison and Warren Franscioni, caught wind of this phenomenon and pounced. They began to manufacture a plastic version of the device, which was much more aerodynamic and a lot more fun. The Morrison/Franscioni partnership did not last long though.

Morrison went on to produce another version called the 'Pluto Platter' hoping to cash in on the recent UFO craze and the nation's fascination with space. The Pluto Platter caught the eye of Richard Knerr and Arthur 'Spud' Mellin at their new company Wham-O, which had just released the Hula Hoop and the Super Ball. Upon hearing the story of the saucer's origins, Knerr changed the spelling of 'Frisbie' and trademarked the flying disc. It's been known as the Frisbee ever since. And it all started with a procrastinator, a baker and a Hula Hoop maker...

Scared of rugby?

Ultimate Frisbee is the ultimate team game for the less robust members of society. Two teams of seven over-excited IT workers/management consultants/accountants compete to throw a Frisbee from man to man towards an end zone without the opposition intercepting the disc in flight. Yay. If you overhear a player mention 'Ultimate', it's not because they're trying to hide the Frisbee aspect. Since a Frisbee is a trademark object, the sport is officially called just 'Ultimate'. There are over 5 million players in the USA, and at least 14 in the UK.

Drum

TAGLINE: anything upon which one bangs
PREDECESSOR: clapping
LESSON: tap into an industry that's booming

> ### Did you know?
>
> Any instrument that is played by beating on a membrane stretched across a hollowed-out body is called a membranophone.

The concept of the drum is as old (and annoying) as humans banging their hands on any hard surface. It seems we as a species have an inherent need to make noise in order to attract attention to ourselves. So there's really no one person who invented the concept of 'drumming'; however, the drum as we know it has evolved quite a bit.

As you can tell, there's a difference between drumming on the ground and drumming on something that's been hollowed out, like a log, a gourd or a head. The idea of using empty space to amplify sound is the basis behind the membranophone – any instrument that is played by beating on a membrane stretched across a hollow body.

It's presumed that the original membranophone was built from a hollow tree stump with an animal skin stretched over it. A little different from the kit gathering dust in your attic, but structurally it's the same idea.

Today, drums are made from a variety of materials. The idea of the 'drum set' has also changed. Rather than a stump and a couple of gourds, the modern drum set typically includes a bass drum, a snare drum, tom-toms, a floor tom and a cymbal (with the option to go overboard and add more variations to your kit). But while the construction has changed over time, the purpose remains the same. The drum is a means of communication.

It's used in just about every musical genre from rock to hip-hop and has cultural variations like the Cuban bongo and African conga. But spare a thought for the humble drum-maker. It's actually a risky job – in the past thirty years at least six craftsmen have died after coming into contact with the spores of anthrax. That's not some new-wave extreme death-metal band, but a disease that can be found on the animal hides used in the making of some types of drum skin.

Golf

TAGLINE: strolling about aggravated

PREDECESSOR: strolling about un-aggravated

LESSON: necessity's not always the mother of invention... in this case, it's not even a third cousin.

Have you ever whacked a rock with a stick? Well, then congratulations, you could have invented golf.

In fact, this pastime is so simple that it was invented a full three centuries before mankind entertained the idea that the earth revolved around the sun. And, like so many other inventions, it appears to be the direct result of boredom. Details are a bit sketchy, but as far as historians can tell, around the mid-1300s, the advent of golf went something like this...

Bored Scotsman #1: Hey, Angus, I'll bet you a sheep's bladder I can get this rock into that wee hole over there using only my staff, in fewer tries than it takes you to do the same.

Bored Scotsman #2: Why the hell would I want to do that, Hamish?

Bored Scotsman #1: Beats having to stand around in the wind and rain taking orders from Mel Gibson.

They then proceeded to do just that until dusk. And that, more or less, is how you begin and end a game of golf today, give or take a sheep's entrail. By 1447, the game became so popular that Scottish men were neglecting archery practice – not a good idea when the English decided to go 'on tour' around the kingdom's borders again. For this reason, King James II banned the game for fear of invasion. (This is the first recorded contact that golf had with Nike... though not the shoe company, the actual Goddess of Victory.) Seriously though, within 100 years golf was so popular it got the entire Kingdom of Scotland in trouble with their boss and it's been doing it with golfers ever since.

You're breakin' my balls...

North Korean 'Dear Leader' Kim Jong-il is one of the world's weirdest enigmas. A known DVD collector who ordered the kidnapping of a South Korean film director to make a *Godzilla* rip-off for him, his life is surrounded by self-generated propaganda, mystery and outright lies. One of the most classic examples is the legend that, according to state propaganda, on his first ever 18 holes on a golf course Kim shot a world-record 38 under par including 11 consecutive holes-in-one. Yeah, 'course he did. He then decided not to ruin the game for everyone else by immediately announcing his retirement. What a considerate fellow.

Football

TAGLINE: twenty-two people trying to kick a ball into a net during a riot

PREDECESSOR: less emotional ball kicking

LESSON: boredom is the stepmother of invention

The game of football can be traced back to a number of different ancient cultures (surprisingly, none of them Brazilian). The Romans, Greeks, Chinese and medieval Europeans can all stake a claim to the ball-in-net pastime.

Archaeological evidence proves a game with similar construct was played way back when by the multi-deity-fearing Romans and Greeks. Putting a military spin on the sport of Pelé, soldiers in Han Dynasty China built their skills by kicking a ball into a net. But it took the medieval English to give us the riot-inciting game we love today.

Way before Watford versus Luton Town, medieval villages squared off in football-like matches and started bitter rivalries with each other. As in today's version, kicking, punching, diving, swearing, biting and gouging were allowed. Hundreds of townspeople came out en masse to take part in the games and accompanying hooliganism, which could go on for hours upon hours – even up to a whole day. There are even rumours that a decapitated human head was sometimes used as the ball. These matches and rivalries proved so violent that the feudal government tried to ban the sport completely.

In 1333, King Edward III passed a number of laws in an attempt to stop the matchday madness. Scotland's King James I followed Edward's lead in 1424 by decrying, 'That nae man play at the Futeball!' It didn't work.

Today football is the most popular sport in the world. Whereas 'soccer' isn't even as popular as midget gator wrestling in the US, despite 'Becks appeal'.

Did you know?

Last year the English Premiership, widely perceived to be the best football league in the world, generated revenues just shy of £1 billion – almost a quarter of all football-related revenue generated in Europe. There are 53 member associations and leagues in UEFA. You do the maths.

Fishing Rod

TAGLINE: a stick, string and hook attached to a guy with wishes of gullible fishes

PREDECESSOR: less suspicious fishes (say that quickly ten times)

LESSON: give a man a fish; you have fed him for a day. Teach a man to fish; you can sell him fishing equipment and beer

Sure, in the old days fishermen had nets and spears to help in their maritime hunt – but those methods, along with trying to snag a fish with your bare hands, required you to go to the fish. The invention of the fishing rod brought the lazy man's logic to this process. Why go to the fish when the fish could come to you? Brilliant! Judging by the ancient 'Gone Fishing' signs found hanging from doors of prehistoric ruins, the bring-the-fish-to-me mentality spread throughout the old world. Artefacts of angling have been found in Egypt, China, Greece and Italy, as well as England, proving medieval Englishmen knew a thing or two about catching fish even before the deep fryer was invented.

Bad joke alert

Two parrots sitting on a perch. One turns to the other and asks, 'Can you smell fish?'

Modern rods may have a few minor differences from the ones back then (such as the use of graphite poles rather than sticks), but the concept has stayed the same. It's the most convenient and efficient way for a person to gather fish. So man has remained fed and happy due to his ingenuity, while fish, on the other hand, have remained pissed off ever since.

Aside from the fact that the fishing rod has helped to feed billions of people since its invention, it's helped to make man a boatload of money. In the UK alone, fishing is a £700m industry. Looking for that next big idea? Go fishing.

Barbie Doll

TAGLINE: unrealistically skinny adult doll for children
PREDECESSOR: crappy German doll
LESSON: girls like to play 'dress up'

The doll is believed to be the oldest toy on record. And the Barbie doll is the most popular doll in the history of dolldom. And it's all thanks to Ruth Handler.

She invented the busty blonde in 1959 and named it after her daughter Barbara. Her intent was to make the doll a teenage fashion figure.

It all started when Ruth took a trip to Europe and came across the German Bild Lilli doll. It was exactly what she had in mind. She bought three dolls, giving one to Barbara and brought the other two home to show her husband, an executive at Mattel.

Controversial figures

If Barbie were a real woman her measurements would be 36-18-38. Impressive, but impossible. Her outrageous dimensions continue to cause controversy. The fact that the Ken doll has no penis doesn't seem to bother anyone though.

She then reworked the design of the German doll and pitched the Barbie doll to Mattel. They loved it. And upon its release, little girls across the United States loved it too. (More than 350,000 Barbies were sold during the first year of production alone.) In a move made to solidify Barbie's worldwide domination, Mattel bought the rights to her predecessor, Bild Lilli, in 1964. And rather than bring the German doll into Barbie's fold, the company ceased its production. They stiffed Lilli. With Lilli out of the way, Mattel has gone on to sell more than a billion Barbies worldwide.

Skateboard

TAGLINE: boards for, like, bored surfers
PREDECESSOR: like not
LESSON: offer alternatives to existing fads

Dudes, even though nobody, like, for sure knows who that one stoked guy was who totally got the skateboard thing going, it's like a pretty solid guess it went down like this...

California Surfer Dude #1: Bummer, dude. There's totally no swells today.

California Surfer Dude #2: Major bummer, dude. It's completely glassed over.

California Surfer Dude #1: You know what'd be an extra-gnarly alternative, bro?

California Surfer Dude #2: Sticking some roller skate wheels to a board and going down a major hill?

Like, happy birthday skater dude

The sport eventually became so popular in the US that a 'Go Skateboarding Day' was officially named in 2004. It's celebrated each year on 21 June. Which is totally rad.

California Surfer Dude #1: Uh, I was gonna say get high, but that totally works, man.

California Surfer Dude #2: Stellar. But let's get high first.

Okay, so maybe that's not exactly the way things went, but you get the picture. The first skateboards were sold in Los Angeles at a surf shop, as a non-aquatic activity for California surfers. It began as the brainchild of Bill Richard, owner of the Val Surf Shop, who struck a deal with the Chicago Roller Skate Company to provide wheels for Richard's boards. And thus, 'sidewalk surfing', as it was originally called, was born.

Bell-Ringing

TAGLINE: a great gong that goes ding, dong
PREDECESSOR: not gonging
LESSON: make making music simple

Ringing a bell – what could be a simpler or more joyous way to make music? (Okay, by its inclusion in this book, 'banging a drum' would be an appropriate response.) Like a drum, a bell can be made from just about anything. Some are clay, some are wood, but most modern bells are usually metal. In terms of where it hangs in the musical family, the bell is classified as a percussion instrument.

And while its simplicity makes it nearly impossible to identify when exactly the first bell was rung, we do know that it is one of the oldest musical instruments ever invented. Bells are depicted in stone tablets and other carvings that date to the fourth century BC.

Whether they are used as part of religious rites or as a mere decorative piece on a mantel, one thing is for sure: bells are, have been and always will be a huge part of human culture. But they're not just used for music-making or time-keeping or calls to prayer. From before the Spanish Armada in 1588 and as late as the 1940s, the bells in English parish churches formed a vital link in the national defence in case of the threat of invasion from the Continent. Each parish would ring its bells until the alarm was spread as far as there were churches to receive and pass on the message. Conversely, what better way to signal a famous victory than to unleash a national pealing of bells? The bell proves you don't need all sorts of bells and whistles to create a powerful sound invention.

Famous bells

Believe or not, there are some famous bells around the world. Here's the lowdown on five important ones.
- Big Ben – the 13.5-ton bell inside the Great Clock in the Clock Tower at the Palace of Westminster, aka the Houses of Parliament in London, one of the most iconic structures in the world.
- The Liberty Bell – located in Philadelphia, Pennsylvania, it's a symbol of American freedom, having been rung on 4 July 1776, the original Independence Day.
- The Great Mingun Bell – at 90 tons it's the largest functioning bell in existence, and it's found in Burma.
- The World Peace Bell – located in Newport, Kentucky, at 33 tons it's the largest swinging bell in existence.
- The Tsar Bell – the absolute largest bell still in existence, at 216 tons, but never rung as a piece fell off during casting; it's on display in Moscow.

HOME AND OFFICE ESSENTIALS

Post-it Notes

TAGLINE: a scrap of paper and some glue
PREDECESSOR: a scrap of paper and some tape
LESSON: the solution to your problem might be the solution to everybody's problem

It's the summer of 1974… you've embarrassed yourself copying the chopsocky dance moves to 'Kung Fu Fighting', watched Teutonic efficiency deliver the World Cup to the West Germans on home soil, and just tested your sister's brand-new Magna Doodle into an early grave.

Meanwhile, 3M employee Art Fry is trying to figure out how to prevent his bookmark from falling out of his hymnal during choir practice. His colleague, Dr Spencer Silver, has just developed a new adhesive that's sticky, but not too sticky. It allows users to fasten things without leaving behind residue, making it possible to reposition once-stuck things. A problem-and-solution match made in inventor's heaven.

Fry used Silver's glue to develop a solution to his problem. Soon he was belting out 'Ave Maria' without a care in the world, his place firmly marked by a semi-sticky little note. Perhaps a little slow on the uptake, 3M didn't introduce the Post-It Notes until 1977 and when they did, people just didn't get the idea and the product floundered. But persistence paid off. One year later, 3M inundated a market in Boise, Idaho with free samples. After trying the notes, nine out of ten people said they'd purchase the product... if only they had a little note affixed somewhere to remind them. A decade or so after their introduction, Post-It Notes were stuck everywhere.

The little notes quickly changed people's organizational behaviour. In the 1990s, Post-It Notes sales reached $15 billion. To date, 3M has sold an excess of 1 trillion sticky notes. Widely considered one of the most important inventions of the twentieth century, they have even been displayed in New York's world-famous Museum of Modern Art.

Fry has since received many accolades for his role in revolutionizing office communication. Today, he travels the world to speak on the topics of creativity and innovation. And you? Well, I'll bet you got a big kick out of the 'artistic' anatomy-related doodles you scribbled on your sister's toy before you broke it. Yeah. Very clever.

> In the 1990s, Post-It note sales reached $15,000,000,000. Which reminds me, I need to make a note to buy some more. D'oh!

Duct Tape

TAGLINE: super sticky silver tape

PREDECESSOR: repairing things properly

LESSON: if your product can hold the universe together, you're on the right track

Jury rigging, or Jerry rigging as it's sometimes known, has been in existence since prehistoric man hastily repaired his woolly mammoth-trunk showerhead with the very first roll of duct tape... oh wait – that was a Flintstones episode. Nevertheless, man has made an art form out of the quick fix and the invention of duct tape brought him a brand new way of hastily slapping things together.

The story of duct tape starts with the invention of adhesive – the sticky in the super sticky silver tape. This sticky substance is a complicated chemical concoction and therefore not included as an item in this book. However, its application in the form of duct tape is simplistic brilliance.

The first people to turn the adhesive into the duct tape we know and love today were the sticky-fingered hotshots in the Johnson and Johnson Permacel Division in 1942. While duct tape's original use was to keep moisture out of ammunition cases during the Second World War, military personnel found it to be a quick fix for just about any problem. Soldiers used it on everything from their artillery to their vehicles. After the war ended, the decision was made to change the tape's colour from Army green to civilian silver since it was intended for use on metal ducts. Although, just as soldiers did during the war, people found that its strength, versatility and durability made it useful for all sorts of things throughout the home.

Today, duct tape is one of the most widely sold adhesive products in the world and remains an essential part of any home tool kit.

So is it *duck* tape or *duct* tape?

It's both. Because of its waterproof qualities, military personnel referred to the tape as 'duck' tape. After the war, construction workers used it on air-conditioning and heating ducts. Hence the slight change in the name from duck to duct.

Tin Opener

TAGLINE: a quick fix when you want to fix dinner quick
PREDECESSOR: chisel and hammer
LESSON: think ahead

The invention of the tin opener is a story of thick metal and even thicker heads. You would think the creation of the tin opener would chime with the creation of tinned food. You would think that... but you'd be wrong.

Cockney merchant adventurer Peter Durand patented the concept of tinning food in 1810. And he used the mighty Royal Navy to help test the longevity of the tinned food's edibility. Durand was all 'Oi guv, check out my tinned fruit cocktail', and the Navy stocked up and set sail. But it wasn't until they were in the middle of the ocean that they were like, 'Blimey! How in the bloody hell do we open these things?'* See, Durand worked out how to seal stuff in really well, but he didn't really think through how to remove it. His tins' instructions read something like 'Cut round the top near the outer edge with a chisel and hammer'. Unfortunately an American inventor takes the credit for taking things one step further. Ezra Warner received a patent for his can opener in 1858. The invention resembled a bent bayonet and was used by first puncturing the can with its point and then dragging it around the rim. A little more work than the one you have in your kitchen today but almost identical to the bit on your Swiss army knife that you never use.

Our modern tin opener (the one with the crank and wheel that does most of the work for us) wasn't invented until 1870. It too was invented by an American, William Lyman. He made it possible for the novice to use the device, which led to mass consumption that, of course, led to one rich Mr Lyman.

So, the modern tin opener was patented 60 years after tinned food. Good thing it keeps, eh?

> I've taken just a little bit of artistic licence in recounting how the Durand/Royal Navy tin-opener relationship played out.

Paper Towel

TAGLINE: disposable absorbent cloth
PREDECESSOR: non-disposable absorbent cloth
LESSON: convenience is king

The paper towel! Perhaps the most wasteful household item in the history of mankind. Before this whole 'save the earth' business got started, I would rip off a paper towel and throw it in the bin each time I walked by the roll for no reason whatsoever other than sheer force of habit. Nevertheless it has made quite an impact over the years.

The history of paper towels all starts with a train car full of toilet paper. You see, a trainload of bog roll was on its way out from the Scott Paper Company in 1907 when the company president received some unpleasant news. The paper in the railcar was too thick to be used as toilet paper. (I can't make this stuff up... well, most of it, anyway.) Now, not to provide too much personal information, but I can use just about any thickness when the need arises. So I don't know what kind of alarmist gave the chief that call.

> When the Scott Paper Company first manufactured toilet paper, it declined to put its company name anywhere on the product's packaging.

But anyway, the president had a plan.

Determined to prevent this thickness issue from hurting his bottom line, he instructed his people to cut the too-thick tissue into larger sections for non-arse related wiping. And thus the Scott Sani-Towel was born, marketed as a more hygienic alternative to the germ-filled cloth rags people were using to clean their kitchens. You'd think they'd fly off the shelves, but it seems people back then weren't too concerned about spreading salmonella all over their countertops. It wasn't until 1931 when the perforated version we know and love today was introduced that people started paying attention to the product. These days paper towels are marketed by stubbly transvestites, dinosaurs and badly-dubbed, unconvincingly clumsy European kids, all with one aim in mind: to make you feel better about your poor domestic hygiene.

Toilet Bowl

TAGLINE: bowl in which one defecates

PREDECESSOR: everywhere else

LESSON: you gotta do what you gotta do, so make it easier for people to do it

Lavatory, can, throne, bog, John and so on and so on – the number of names for this extremely simple device suggests only one thing: no one actually wants to say what it's for. And for the purposes of this book I'm not going to say it's a receptacle for the efficient disposal of adult human faeces either.

Like any good invention, there's some disagreement over who should be credited with the invention of the modern toilet. Though the controversy continues to swirl about (clockwise in most civilized societies... anticlockwise in Australia), I've flushed out two main contenders, both of whom are British. Most experts in these gritty matters agree that J F Brondel introduced the first valve-type flush toilet in 1738. However the Elizabethan gentleman Sir John Harrington is credited with inventing the 'water closet' in 1596. Showing a fine sense of humour, he outlined his ideas in a pamphlet titled 'A New Discourse on a Stale Subject' and went on to design a loo with a flushing tank and a mechanism for washing out the toilet bowl. Elizabeth I, his godmother, is rumoured to have had one installed but fought shy of using it because of the racket it made when operated. Seems the British love affair with all things toilet-related wasn't matched elsewhere: the first American foray into the world of flushing toilets didn't come until 1857.

Thus, despite what many people think, Sir Thomas Crapper did not invent the toilet. However, Thomas Crapper was issued nine patents; four were for drain improvements, three for 'water closets', one for manhole covers and one for pipe joints. So, if you think about it, it's not a huge leap to associate Tom with the word 'crap', even if it's a load of.

A toilet by any other name still smells

If the word 'toilet' just isn't titillating enough for you, there is no other shortage of other names for it... privy, loo, washroom, commode, lavatory, shitter, porcelain throne, potty, dump station, black hole, rest room, powder room, john, lav, little boys' room, water closet, bog, the reading room, turd pool, 'the facilities', public convenience, thunderbox, and so on.

Toothpick

TAGLINE: a little pointy stick
PREDECESSOR: loud sucking noises
LESSON: make it easier for people to do something they already do

Food has been getting stuck in people's teeth for over 400,000 years. Neanderthal and Homo sapiens fossils show clear signs that even early humans picked their teeth with rudimentary tools. So you'd think some caveman would have opened up a little toothpick emporium way back in the day.

Lord knows if the Flintstones taught us anything, it's that prehistoric man was a sucker for get-rich-quick schemes. But it wasn't until 1858 that somebody figured out how to take toothpicks all the way to the bank.

Charles Forster of Strong, Maine, is believed to be the first person to manufacture toothpicks. And it wasn't as easy as you'd think. He had to go all the way to South America to work it out. In his travels, Forster watched natives use little pieces of wood to clean their teeth. Impressed by the technology that had been under our collective noses for 400 millennia, he sent a sample box home to his wife who showed them around. 'Look at these!' she said to the people of Maine, who stared back at her dumbly and made sucking noises in an effort to dislodge the food from their teeth. After a quick and disgusting demonstration, Mr Forster had more orders than you could shake a tiny little stick at – especially from hotels and restaurants.

Forster's first few batches were handmade; the Model T of toothpicks you might say. But by 1860, he had to devise machines to keep up with the growing demand. The genius of the toothpick-making machine was that it allowed small sticks to be cut into, er, much smaller sticks.

Today, there are toothpicks-a-plenty in nearly every household and eating establishment in the world. Their dental applications aside, without them our club sandwiches would be completely unmanageable, our fingers would be all over those free samples at ASDA and our tropical drinks would be lacklustre in the absence of tiny paper umbrellas.

Did you know?

Early Romans used porcupine quills as toothpicks. Late Romans found there were no clean ones left.

Paperclip

TAGLINE: the staple for people with commitment issues
PREDECESSOR: a straight pin stuck into stacks of paper
LESSON: most people are unorganized; organize them

We have all reverse-engineered a paper clip at one point in our lives, only to discover that our suspicions were correct: it's a wire.

Wire was invented as early as 1500 BC, so why did it take so long to make this mental leap? Lots of reasons... steel wire was still new, machinery had to be introduced to manufacture it and so on. But mainly, it's because you, or someone like you, didn't think of it until the 1860s when canny American Samuel B Fay made millions from the idea. It's this type of tiny, cricket-sized mental leap that separates the everyday entrepreneur from the billionaire. All you have to do is get out your tiny mental cricket and give him a little shove.

> Over 100,000,000,000,000 paper clips have been sold. About the same number are now permanently 'on loan' from the stationery cupboard.

Fay patented the paper clip in 1867. The original patent listed its primary purpose as attaching tickets to fabric, but it did make mention that it could be used to organize papers. A subsequent flood of paper clip patents was issued beginning in 1868 as hundreds of would-be inventors were audibly heard striking their foreheads and saying, 'D-oh!' Nowadays you can't sit at your desk without seeing at least one paper clip, which to date have sold over 100 trillion units, though how anybody keeps track is beyond me. It's amazing to think that people have purchased that many pieces of bent wire. But it goes to show that a great idea will go a great distance.

'Many of life's failures are people who did not realize how close they were to success when they gave up.'
- Thomas Edison

Wire Coat Hanger

TAGLINE: mangled and twisted piece of wire
PREDECESSOR: wooden hangers
LESSON: be careful. Protect yourself and your idea

At a pinch, they help us break into cars, unclog our toilets and faithfully serve as old car radio and television antennas. But, mainly, they provide an inexpensive hanging place for nearly any article of clothing. And it's all thanks to a bloke called Albert Parkhouse.

Like every other morning, Albert arrived to work one day in 1903 at the Timberlake Wire and Novelty Company, a company in Michigan, USA, that made lampshade frames and other wire items. Flush with excitement at the prospect of making lampshade skeletons and other wire novelties for painfully low wages, he went to hang his hat and coat on the hooks considerately provided by the company. But all the hooks were in use.

> 'Don't worry about people stealing your ideas. If your ideas are any good, you'll have to ram them down people's throats.'
> - Howard Aiken

He thought about just putting his coat back on, but the prospect of getting a sleeve caught in the machinery convinced him otherwise. After all, maimed workers were sent home early, without pay and occasionally without limbs. Ever the dedicated worker-bee, Parkhouse picked up a piece of wire, bent it into two large hoops and twisted both ends at the middle to create a hook. He hung up his coat and began his working day.

What did Albert get for his vigilance? His boss, Mr Timberlake, happened to pass by, no doubt on his way to count his money, and promptly took out a patent on the 'hanger'. The company made a fortune and took credit for inventing one of the most useful devices in modern history.

What about Parkhouse? He died penniless, but at least his shirts were crease-free.

Rubber Band

TAGLINE: stretchy little belt of rubber
PREDECESSOR: boring, un-shootable string
LESSON: boing!

British inventor Stephen Perry of the London-based company Messrs Perry & Co patented the rubber band in 1845. But it took a newspaper to really popularize the invention.

It's probably no surprise that a newspaper had a hand in popularizing the rubber band. But it's probably a surprise that it wasn't because the newspaper wrote about the product. Nope. The PR came from the paper using the band so its paper-boy deliveries didn't blow all over the neighbourhood.

On 7 March, 1923, William Spencer of Ohio got hold of some rejected inner tubes from the Goodyear Tyre & Rubber Company and began cutting them into bands in his basement. Spencer, who worked for the Pennsylvania Railroad, began trying to sell his rubber bands to office-supply stores and paper and twine outlets.

One day he noticed the indispensable *Akron Beacon Journal* blowing across his lawn in bits and persuaded the *Beacon* to bind their papers with his rubber bands. He talked the riveting *Tulsa World* into doing the same and persuaded grocers to use his rubber band instead of string to secure produce.

Spencer continued working for the railroad for fourteen years while building his rubber-band business. By 1944 he was able to open a second plant in Arkansas. In 1957, he opened another in Kentucky and expanded to California in 1988. The Alliance Rubber Co. now produces more than a million kilos of rubber bands a month.

Nowadays we often see red rubber bands instead of the traditional brown ones. This is because in 2004, after complaints from the public about postmen littering their mail rounds with discarded rubber bands, the Royal Mail decided to start using a red variety which are easier to spot on the ground. The theory was that idle postmen would pick them up and not be tempted to drop their own. But the plan backfired, and customers moaned about an 'epidemic' of red elastic bands instead. One local politician talked of finding 37 in one street in Colchester. With over 350 million in use every year, the Royal Mail can spare a few – which is why they sent 20,000 to be used in the world's biggest elastic band ball. But instead of bouncing spectacularly high into the air on a test range in Arizona, the ball gouged a massive crater in the desert after being dropped from the sky at half the speed of sound.

How far?

A large-ish rubber band of 6mm thickness, 80mm diameter will stretch to up to six times its length without breaking, given the right temperature conditions.

Pencil

TAGLINE: the poor man's pen
PREDECESSOR: charcoal
LESSON: the pencil is greater than the sword

Think that's really lead in your pencil? Think again. Lead has never been used in pencils. It's all about graphite. Graphite became a popular choice when a large chunk of it was discovered in England nearly 450 years ago after a storm uprooted trees in the Lake District (home now to the excellent Cumberland Pencil Museum). This source of graphite (literally from the Greek 'to write') remains the only one ever to have been found in such a pure and solid form. The discoverers thought they'd found lead, hence 'lead pencils'. Over two hundred years later, an English scientist worked out that the substance wasn't lead after all. While it left a better, darker

Did you know?

The world's biggest pencil is a Castell 9000, housed in a glass tower outside German pencil firm Faber-Castell's plant in Kuala Lumpur, Malaysia. It is 65 feet tall and took two years to make.

mark than lead, and proved handy for marking local sheep (hopefully not out of ten), graphite was too soft to be used on its own. After mucking about with different forms of protective coating, including sheepskin and string, eventually the coloured casings we know and love and chew today came into play courtesy of English and Italian inventors. Pencil makers began hollowing out cedar sticks and inserting the graphite, creating a graphite-fuelled writing instrument that didn't break in your hand. Mass production followed and today over 13 billion pencils are manufactured worldwide – enough to circle the Earth 62 times over.

Paper Cup

TAGLINE: cups you chuck
PREDECESSOR: the no-chuck cup
LESSON: convenience is king

'Death in School Drinking Cups!' was the title of a study by Alvin Davison, an American biology professor. The paper was published in the August 1908 issue of *Technical World Magazine* and was based on research done in the Pennsylvania schools system.

Davison was so alarmed because during the early twentieth century, it was common to have shared cups at public water sources. As you can imagine, the sharing made people ill. Everybody was diving into the same stagnant pools like a bunch of pigs at a trough! Luckily, Davison's study came out the same year as the paper cup was invented. So his call for alarm paired with the availability

> 'Shun idleness. It is a rust that attaches itself to the most brilliant of metals.'
> - Victor Hugo

of disposable cups resulted in the passage of many laws ending the use of shared drinking cups in schools.

It took a little longer for hospitals to catch on to the paper-cup craze. It wasn't until 1942 that an east coast university published a study exposing the high-cost of cleaning glassware for reuse in hospitals. As it turned out, the cost of reusing glasses was nearly two times as high as using the disposable alternative. A loss for Mother Nature, but a win for people going into the hospital and not leaving dead. And a win for all of us who use vending machines and burger vans, or who can't be bothered with washing up. In short, lazy, unhealthy specimens most likely to have need of a hospital in the future.

Masking Tape

TAGLINE: a painter's best friend
PREDECESSOR: stickier tape and butcher paper
LESSON: trust hard-working employees with initiative

The 3M Company is mentioned several times in this book and it seems like the story is always the same. It goes something like this:

Innovative 3M Employee: Hey! I've got this great idea for a new product!

3M Manager: Get back to work on that other sticky thing.

Innovative 3M Employee: Look!

3M Manager: Get back to work on that other sticky thing.

Innovative 3M Employee: You sure about that, boss?

3M Manager: Ah... maybe you're right. Let's make billions of dollars off your idea. Now get back to work on that other sticky thing.

And such was the case with masking tape.

> 'Try and fail, but don't fail to try.'
> - Stephen Kaggwa

One afternoon 3M employee Dick Drew went to a car repair garage to test some new sandpaper he was working on. But rather than get feedback on the sandpaper, he received an earful about painting. It seems the mechanics could no longer deal with the tape and paper combo that was used for two-tone paint jobs. It just didn't do the job. In their frustration, Drew saw opportunity. He returned to his 3M lab determined to help those men out. He set out to create a paper/tape hybrid that would solve their coverage issue. And he did just that. But not before he got in trouble with then-company-president William McKnight. It seems McKnight didn't want Drew spending time on such a project. Luckily for him, 3M and us, Drew didn't listen. He kept at it and invented masking tape, a sort of sticky-paper that could be used to block off areas a painter didn't want to paint. (Of course people found other uses for the tape as well.) Goes to show – even if it gets sticky at times, it's best to persevere.

Drawing Pin

TAGLINE: a miniature nail

PREDECESSOR: the nail

LESSON: make a 'light' version of something that's already useful

The modern drawing pin was derived from Edwin Moore's map pin. (Map pin? Ever gone on to a website to get directions and used those virtual pushpins to determine your location? Those, except real.) In 1900, Moore founded the Moore Push-Pin Company with a hundred dollars and as the founder and only employee, Moore spent all his time laboriously producing these pointy pushpins.

The first sale was for one gross, or twelve dozen, that brought in a whopping $2.

Orders went up from there though as giant US photo company Eastman Kodak soon placed one for $1,000 (approximately $25,000 nowadays). The company used Moore's invention to 'tack' up their photographs.

And that brings us to how Moore's map pushpin evolved into the drawing pin. In 1904, a few years after Moore founded the Moore Push-Pin Company, German clockmaker Johann Kirsten designed the first drawing pin. Similar in construction, Kirsten's pin had a flat head opposite the point, whereas Moore's pushpin had a tiny grip of sorts.

Besides the flathead versus grip difference, the two men also went in different directions with their lives post-creation of the pointy inventions. Moore went on to invent and patent a number of other items, including the picture hanger. (Perhaps he had a hang-up for hanging stuff up.) Kirsten on the other hand sold the rights to his invention to Otto Lindstedt, who patented it in 1904. Lindstedt became very wealthy. Kirsten remained broke. Not too sharp – the drawing pin remains one of the top-selling office products of all time.

Carrier Bag

TAGLINE: a disposable sack

PREDECESSOR: bringing a non-disposable bag with you (heaven forbid!)

LESSON: today, disposable items are no longer an option; create an item that does the same thing without harming the environment. Good luck with that.

Imagine a time long ago when people took their own canvas bags to the supermarket to carry home their shopping... What's that? The time 'long ago' was last Sunday? Oh, right. Well, imagine it's a few years ago – before the global warming agenda, Al Gore and his *Inconvenient Truth* – now, try again to imagine a time long ago when people took their own bags to the shops.

An outrageous idea, yeah? Why would someone bring his own bags when there are all those free carrier bags to use? The plastic bag was the way to transport your milk, crisps, frozen chicken nuggets and other incidental purchases before it became environmentally uncool to do so. The first carrier bags were introduced in the 1970s and these days over a trillion are in use around the world each year. Apparently that's enough for every human being to own 179 of the bloody things. And I bet you hoard them at home but never re-use more than a few at a time, necessitating bringing home even more bags every time you pay a visit to the shops.

It's been calculated that four out of five of every carrier bag dished out in the US is made of plastic. Which leaves the other 20 per cent to the paper bag.

The first person to make any money out of the paper bag was Francis Wolle. Not only did he invent the paper bag, he invented the machine that mass-produced these throwaway carriers. And in 1869, he founded the Union Paper Bag Machine Company with his brother and other partners looking to cash in on the paper bag craze. (Okay, maybe 'craze' is a bit of an overkill – but it was well thrillin' for Wolle and his posse.) However, his bags aren't the kind you might see today in American shopping malls. No, the square-bottomed bag you see was invented by Margaret Knight. You could say inventing was her 'bag' as she's the noted inventor of ninety creations and holds twenty-two patents. So it's the square-bottomed old bag who can stake a real claim to fame having made millions from her invention.

Old bags

It's estimated that about 13 billion plastic bags are handed out each year in the UK alone. Each bag is probably used for an average of 15 minutes but they take 1000 years to decompose.

Pillow

TAGLINE: a soft place to rest your weary head
PREDECESSOR: harder places to put your head
LESSON: where heads are concerned, softer is better

What could be more common than a pillow? Everyone has several, sometimes dozens, even hundreds! (Okay, so not hundreds – but a lot.) You probably can't even imagine trying to sleep without a soft cushion under your head. But that wasn't always the case.

People living in ancient Egypt really lived the hard-knock life. For many, lying down to go to bed meant resting their heads on pieces of wood or slabs of metal. (No wonder they were always at war – they were grumpy from a bad night's sleep.) Back then they believed soft pillows would make it easier for demons to possess their sleeping bodies. So they chose the annoyance of uncomfortable rest over the need for an exorcism.

It took the ingenuity and infamous decadence of the ancient Romans and Greeks to transform the pillow fight into a playful event – as opposed to the bludgeoning bloodbath it was with those Egyptian pillows. These two civilizations went for the softer-is-better approach, resting their heads on straw and feathers. This idea caught on. Soft pillows and cushions became a sign of prosperity. As the years went by, pillows became fluffier and those wood and metal Egyptian versions turned into a distant nightmare.

The sale of soft pillows is now a multi-million pound industry. The sale of hard ones is not.

Pillow fights

Not just a staple of childhood sleepovers or pyjama parties, pillow fights have had a renaissance in recent years. They are the subject of a bar-based Pillow Fighting League in Canadian cities, form the basis of a female wrestling format in the World Wrestling Entertainment promotion, and have been fought by huge crowds as part of semi-spontaneous flash mob culture. Feather pillows make for more fun, because they split more easily than synthetic fibre varieties.

Matches

TAGLINE: sticks that light on fire

PREDECESSOR: lighting sticks on fire

LESSON: make the ignition of controlled fire easier and it will sell

Here's another example of it taking for-ev-er to put two seemingly obvious things together to come up with a good idea.

Fire is one of the most basic and essential human needs. But before the match, starting a fire was not that simple. Even though one of the main components of the match – phosphorous – was discovered in 1669, it took until 1827 for the first practical match to be invented. It came about when John Walker, an English chemist and apothecary, coated the end of a stick with the material and realized he could quickly light the stick on fire by striking it against a rough surface. Convenient, but toxic – Walker's match let off poisonous fumes.

Two more inventors followed in Walker's footsteps. Samuel Jones came up with his own version he called 'Lucifers'. Unfortunately, these little devils also let off harmful smoke. And then Swede Johan Edvard Lundstrom patented his own version of the fire-starters in 1855. He even separated the phosphorus from the other ingredients by putting it on a sandpaper strip affixed to the outside of the box and the rest of the materials in the match head. Still not safe smoke though. (It'd kill you if you breathed enough of it.) British company Bryant and May took to selling half of Lundstrom's output and cornered the London market using downtrodden 'matchgirls' selling Lucifers. The company found that the public was unwilling to fork out for their safety matches which had been designed to eliminate the effects of 'Phossy Jaw', a disease caused by the harmful smoke. Finally, in 1910, a viable non-poisonous match was made. Using a chemical called sesquisulphide of phosphorous, the American Diamond Match Company managed to achieve what was out of the other three's reach – the non-toxic match, which could be produced at a rate of 600,000 per hour! This was marketed under the famous Swan Vesta brand, and used a formula that is still used in the modern match you light up today.

Taft's tax trick

The reliable ignition of fire was deemed so important that American President William Taft persuaded Diamond Match to release its patent for 'the good of mankind'. Diamond Match complied. And in January 1911 Congress promptly placed a high tax on matches. D-oh!

Toilet Paper

TAGLINE: paper with which one wipes his or her backside

PREDECESSOR: see first paragraph below

LESSON: make it more convenient and, yes, pleasant, for people to do the things that everybody does and nobody talks about

Wool, lace, hemp, hay, rags, grass, leaves, snow, fruit peel, the pages in this book, seashells, clams, moss, wood shavings, seaweed, corn husks, animal furs, sticks, feathers, a baby chicken, dead rats, your under-crackers if you really have no choice, etc – all early substitutes for TP.

And when those weren't used, one's bare hand was always a possible solution. In fact, it was common practice all over the world, from Europe to South America to India, to use the left hand for wiping and the right hand for greeting. Though not usually at the same time.

However, the Romans and Chinese weren't really down with the left-hand wipe. Instead, the Romans used a sponge on a stick

Did you know?

Each year there are more than 40,000 toilet-related injuries in the USA alone.

> **Bad joke alert**
>
> 'What hand do you use to wipe your backside with?'
> 'My right hand – why, what about you?'
> 'I don't – I use toilet paper.'

to clean up after doing their business. And when they were done, the sponge went into a bucket of water to 'clean' it for the next person. It took the Chinese, who invented actual paper, to provide a paper-alternative to all the other cleaning methods.

Americans got in on the act of making toilet tissue in the mid to late 19th century, though modern toilet paper wasn't manufactured until 1907, when the Scott Paper Company rolled it out. The fact that the company's still around today proves how well the public (and their arses) took to the product.

Recycling

TAGLINE: global hand-me-downs
PREDECESSOR: chucking stuff into big pits
LESSON: 'save the Planet' my arse! The planet will be just fine without humans

Think your fancy green recycling bins are something new? Nope. Just ask your grandparents who suffered through the Second World War. They were recycling back then just to get us through the war. Citizens were encouraged to 'dig for victory' to conserve food. There were collection efforts to help with the conservation of iron, tin and copper. So don't think you're so special.

Recycling has been a matter of environmental or economic concern in one form or another since the dawn of man. The only thing that's really changed is that the advent of the industrial age has finally mucked up the environment enough for us to notice. The good news is, we are finally taking action. Recycling

> Not a fan
> Modern technology
> Owes ecology
> An apology
> - Alan Eddison

> **Did you know?**
>
> Battery manufacturer Duracell built portions of its new international headquarters using its own waste materials.

has become a huge and profitable industry, creating more opportunities for small and large companies. But, the conservation of our elements and necessities in order to sustain our lives as human beings has been a consistent element of our fundamental survival.

Today, public officials can sub-contract with privately owned companies, creating new opportunities in the private sector and making recycling one of the fastest growing industries on the planet.

ACCESSORIZE THIS!

Crocs

TAGLINE: cheap, hideous, rubber footwear

PREDECESSOR: ugly shoes that people didn't want to buy

LESSON: it doesn't have to be pretty; it just has to be good

It is said, 'There are no stupid questions.' Sure, it's usually said by someone who wants to make stupid people feel better, but there is truth in that statement. And if you need proof, look no further then the shoe sensation: Crocs.

On a Caribbean sailing trip in May 2002, three allegedly inebriated blokes from Colorado asked themselves, 'If we could conceive of the perfect shoe, what would it be?' Their answer: a shoe they could wear on their boating trips that was comfortable, practical and fun. Taking it a few ugly steps further, they opined that this 'wonder shoe' should be slip-proof, waterproof and not leave scuffmarks and shouldn't smell after getting wet. Always a good rule that one.

When they sobered up it still seemed like a pretty good idea. So, with strictly utilitarian needs in mind, they designed a simple rubber shoe. The shoe was made of non-scuffing durable rubber. Nothing new, really. But then to achieve comfort and aeration they made the shoe wide and roomy and added ventilation holes (which leads one to seriously question their definition of 'waterproof shoe'). They ended up with an extremely good shoe for its designated purpose and in July 2002 they debuted them at a local boat show.

What happened next is the stuff of invention legend. People took to the strange footwear as one might take to a puppy that's so ugly it's cute. Their undeniable ugliness became an instant asset. That coupled with their comfort and usefulness drove demand through the roof and by 2003 they could barely keep up with the deluge of orders.

That's one small step for man...

They've been called 'the ugliest shoes ever made', 'tinker toys on steroids' and 'rubber abominations' (though clearly whoever called them that has never been into the back room of a 'fancy dress' shop). But do the inventors of the Croc mind this criticism? Who knows? They can't stop laughing long enough for anyone to ask them.

Today, Crocs are available all over the world and are one of the most successful shoe stories in the history of footwear. That's one small step for man, one giant, ugly leap for mankind.

Bra

TAGLINE: over the shoulder boulder holders

PREDECESSOR: less support, but happier construction workers

LESSON: create useful garments

For centuries women have had a desire to enhance the shape of their bosom for mostly aesthetic reasons. Actually, I should probably put it this way: for centuries men have wanted to look at cleavage and women have obliged.

The breast-supporting philosophy changed slightly during the nineteenth century when women sought more comfortable underwear than the trussed and squeezed look of those days' corsets. You might say it was a bit of an upheaval. As new underwear creations came on the market, patents were issued at a rapid rate. The US government has granted 1,200 patents for bra-like inventions since the first 'corset substitute' was issued in 1863. Being a 'supporter of breasts' myself, I think that's great.

In 1893 Marie Tucek received a patent for what was literally called the 'breast supporter'. It had two separate cups with straps over the shoulders and was fastened by a hook-and-eye closure in the back. Tucek's patent remains the basis of the modern bra, yet unhooking it still baffles men to this day.

Hold me back

Airbags, bangers, bazookas, boobs, Cannon and Ball, cans, chesticles, chumbawumbas, coconuts, gazongas, hooters, jugs, knockers, lady lumps, melons, milk wagons, norks, puppies, rack, tits, Zeppelins... Call 'em what you want to call 'em, any way you swing 'em, they've been supported by bras since 2500 BC.

Bikini

TAGLINE: wearing your underwear in public
PREDECESSOR: wearing your underwear in private
LESSON: being first isn't everything and sex sells

Talk about the bleedin' obvious. If you were to take every heterosexual male who has ever been born and raise them, individually, in complete isolation, each and every one of them would independently invent the bikini... about three seconds before they invented the naked lady. But it wasn't until 1946 that someone cashed in on the male lust for barely-there, soaking-wet attire and did so with a little help from the atomic bomb.

Don't get me wrong; running around naked at the beach is nothing new. Women have done it for ages. But sadly by 1946, the liberal attitude toward the scantily clad female form had disappeared. Bathing suits looked more like nuns' habits. That's where two Frenchmen come in. Just need a rubber chicken and we'd have the start of a great joke...

Jacques Heim first advertised his two-piece bathing suit over the skies of Cannes, calling it the 'atome' (French for atom) because of how small it was. Three weeks later, Louis Reard unveiled his two-piece number. Using skywriters over the beaches of the Riviera, he proclaimed his suit 'smaller than the

> 'Perhaps imagination is only intelligence having fun.'
> - George Scialabba

smallest bathing suit in the world', and named it the bikini. Reard spun a story about how the name bikini came from the little islands in the South Pacific where the United States had recently tested several nuclear weapons.

The bikini, he said, was named so because he had 'split the atom'. Very clever.

Don't laugh. Take a product that allows women to run around nearly naked, throw them into a post-world war society teaming with sexual tension, add a storyline that links the whole thing to the nuclear bomb, stir in a pun or two or five thousand et voila! You have headlines all over the world and a place in history.

As Heim found out, you can come up with an idea first and still get scooped. Besides, 'Itsy Bitsy Teenie Weenie Yellow Polka Dot Atome?' That wouldn't have worked at all, not even for the cover-version genius that is Timmy Mallett.

Earmuffs

TAGLINE: cold weather cover-ups
PREDECESSOR: cold ears
LESSON: what's good for your ears might be good for everybody's ears

Where else would this invention start but with cold ears? This particular set of ears belonged to Chester Greenwood, a fifteen-year-old with an inventive spirit and a bitter hatred of the bitter cold.

As reported in the *Wall Street Journal*, 'Chester Greenwood's ears were so sensitive that they turned chalky white, beet red and deep blue (in that order) when the mercury dipped.' But you can't believe everything you read (except in this book, of course). I think it's more likely that Chester was sort of a wuss. Most kids would simply put a hat on to solve the problem, but little Chester's ears were sensitive to wool.

So, one day in 1873 Chester decided to do something nerdy about it. He approached his grandmother for advice and help in creating an effective earshielding device. I imagine their conversation went something like this...

Chester: Hey Grannndmaaa! What big cold ears I have!

Grandmother: The better to make millions of dollars with, my dear!

He called their collaborative effort the Greenwood Champion Ear Protector.

And it didn't take Chester too long to create. It was extremely simple – some bent wire, a little insulating material and some sewing. He founded the Chester Greenwood & Company in order to manufacture and sell his invention. According to the official website of the state of Maine, his company had its biggest year in 1936 when it produced 400,000 pairs of earmuffs.

The Ear Protector, or 'muff,' was an instant hit. And it's all because Chester had cold, itchy ears. Lucky he didn't have cold and itchy other parts of his body, or he might've had a harder time getting away with using the word 'muff' on those.

Tie

TAGLINE: useless ribbons of fabric awkwardly fastened to men's necks

PREDECESSOR: freedom

LESSON: Where fashion is concerned, useless is priceless

Men are known even less for accessorizing than they are for their attention to detail. Incredibly this has not stopped us from fastening coloured and often fancily patterned ribbons around our necks for the past 2,000 years.

Throughout human history, men and women have adorned their bodies with neckwear. It all began ages ago when ancient Egyptian men would hang a rectangular piece of cloth around their shoulders as a symbol of social status. The use of the tie evolved over the years and was often used as part of military uniforms. (Presumably to over-accessorize the enemy into submission.) In the late eighteenth century, a looser version of the necktie known as a cravat was introduced into mainstream fashion by an influential group of young men called the Macaronis. These were young Englishmen who, upon returning from the 'Grand Tour' around the continent to countries such as Italy – that's where the 'macaroni' part comes in – brought back new ideas for menswear and fashion.

Have you got a loyalty card, mate?

Tie knots can speak volumes. There are 85 theoretical ways to knot a tie, but most people stick to the main four: the four-in-hand knot, the Pratt knot, the half-Windsor and the Windsor. Contrary to popular belief, the Pratt knot is not the same as the tragically thick one sported by tragically thick footballers, first-time-court-appearance 'clients' and Saturday staff in high-street mobile phone retailers.

So we have those lads to thank for coming to town a-riding on some ponies. Then, presumably, not satisfied with merely sticking feathers in their caps, tying scarves around their necks and calling themselves Macaronis.

Before the Second World War, ties were worn much shorter and wider than today. That's because men in those days often wore their slacks at the waist, rather like music mogul Simon Cowell is said to do. So their ties would only need to hang as low as the middle of their bellies. In the 1960s the 'Kipper tie' became a fashion statement, named in wry honour of its designer, Michael Fish. And you thought it was a whining Brummie version of a herbal brew...

Zip

TAGLINE: toothy fastening device
PREDECESSOR: buttons
LESSON: if your design isn't quite where it should be, keep at it or you'll never know

What's sexier than a woman asking a man (or, hey, another woman) to help zip up the back of her dress? Or better yet, to unzip it? Imagine that same romantic moment being interrupted with the loud rip of Velcro. It just wouldn't be the same. While we have several people to thank for the zip as a mainstay in today's fashion industry, the zip's rise to prominence wasn't exactly a... zippy one. It was a rocky start for the humble zip... galoshes were involved... I'll explain.

It all began in 1851 when one Whitcomb Judson came out with his Clasp Locker device. Judson began marketing his Clasp Locker with businessman Colonel Lewis Walker. The two unveiled their Clasp Locker at the 1893 Chicago World's Fair. It was met with yawns. No one wanted anything to do with Judson and Walker's Locker. In all fairness though, in its original form, it was a pretty scary, medieval-looking thing with big shark-like teeth (seriously).

The two realized something needed to be done. They hired Swedish-born Gideon Sundback, an electrical engineer. He went to work for Judson and Walker's company and redesigned their torturous-looking device into the less-frightening zip we know today.

However, the name 'zip' or 'zipper' was first used by the B F Goodrich Company (yeah, those rubber and tyre guys). They were using Gideon's fastener on a new type of rubber boot they were calling 'galoshes'. In addition to Goodrich's galoshes, tobacco pouches were also using the zip. However, it took twenty more years to convince the fashion industry to seriously promote the zip on clothes.

Today's zips are used on just about everything. They're common on purses, duffle bags, backpacks, raincoats, jackets, coats, jeans, trousers, women's dresses, puppets and, of course, leather masks... erm... so I've read.

Zippy in domestic abuse scandal

The star of classic children's TV programme *Rainbow* was undeniably Zippy. A brown-orange creature of indeterminate species and no discernible grace or mobility, Zippy got his name from the fact his mouth was essentially a gaping horizontal zip. When Zippy's manic-depressive personality and grating drawl became intolerable, his long-suffering chums Bungle and George would zip his mouth up and keep him quiet. The weird thing was, despite having 'hands', Zippy never once unzipped his own mouth to continue haranguing his co-habitees. Was there an unspoken threat hanging over Zippy's head that we weren't aware of?

Velcro

TAGLINE: shockingly loud, tiny hooks and loops
PREDECESSOR: burrs
LESSON: Mother Nature is a clever Mother %*#!-er

One summer day in the late 1940s a Swiss amateur mountaineer named George de Mestral took his dog for a walk in the woods. He and his canine companion came back covered in those prickly little seed carriers known as burrs.

As a part-time inventor, de Mestral was intrigued and inspected one of the annoying things. He saw that the carriers were covered in tiny little hooks that allowed them to cling to things like his dog's fur and the thread of his trousers.

Upon making this observation, de Mestral thrust his face skyward, as lightning struck, illuminating his laboratory in pale blue light and screamed, 'It's alive!' He then set to work designing a two-sided fastener.

One side consisted of stiff, burr-like hooks. The opposite side consisted of loops, much like the threads of his slacks. He called this invention 'Velcro' combining the words velour and crochet. 'It will rival the zip in its ability to fasten,' he announced... presumably to his dog.

Patenting the invention in 1955, de Mestral partnered with a French weaver in order to hone his Velcro fastener. De Mestral was soon selling over sixty million yards of Velcro per year, creating a multibillion-dollar industry which in particular has benefited pensioners all over the world.

So, the next time something inadvertently sticks to you, give it some thought. Maybe there's something you can do with that dog crap on your shoe. I don't know. I'm just saying.

Sunglasses

TAGLINE: glasses you can't really see through
PREDECESSOR: squinting
LESSON: make people more comfortable

The first sunglasses appeared centuries ago when Chinese judges wore smoke-coloured lenses to conceal their eyes during trials. They said it was to prevent people in the court from seeing their expressions during trials. (I think they were just trying to look cool.) From there, English designer and instrument-maker James Ayscough picked up the concept of tinting lenses in the eighteenth century. Though these 'sunglasses' weren't created as a fashion statement either. Instead, Ayscough believed blue-

Beware the Hun in the sun

Today's fighter pilots rely on much more complicated whole-face tinted visors to reduce the blinding effect of the sun at altitude. Can't see those catching on in the same way, sadly. Though don't bet against Beckham or someone donning one for yet another gazillion pounds of tattoo-money.

and green-tinted glass could help correct the eyesight of people with certain vision impairments.

The popularity of sunglasses designed to specifically shade the eye from the sun is really a twentieth-century phenomenon first introduced by the American military. To early military pilots, it was a necessity. The glare from the sun while on flights above cloud-level was extreme and could be the difference between life and death during a dogfight. Thus: Aviator sunglasses were born.

It took some time and a little marketing magic for sunglasses to go from military to mainstream. In 1929, Sam Foster sold the first pair of civilian shades. It didn't take long for them to become all the rage. By 1930, they were the must-have accessory for a sunny day and remain so to this day.

Horseshoe

TAGLINE: equine footwear
PREDECESSOR: barefoot horses
LESSON: look to a beast to ease your burden

A horse may be a horse (of course, of course), but a horse without horseshoes is useless. These very simple iron implements play a very significant role in the transportation history of man- and horse-kind.

The idea of the horseshoe goes way back to early Asia, where horsemen fitted what they called 'horse booties' on their steeds. Taking it a step further, the ancient Romans slipped trendy sandals made of leather and metal on their horses' hooves. However, the Roman hipposandals weren't good enough for European horsemen in the sixth and seventh centuries. (Plus, the horses felt a bit stupid in sandals.) Instead, some horsemen started to nail pieces of metal to the animals' hooves. Things started to take shape during the thirteenth and fourteenth centuries when the widespread manufacturing of iron horseshoes became commonplace.

As for who got lucky when it came to the invention, it wasn't until 1835 that someone really cashed in. That's when Henry Burden created his horseshoe-manufacturing machine, which produced up to sixty shoes an hour. Burden hee-hawed all the way to the bank. And then in 1892, Oscar Brown patented the compound horseshoe that's more along the lines of the shoe used today.

So horses have had a booty, a sandal and a shoe... which means there's still room on the market for a high heel. Get on that!

Feeling lucky?

We all know that horseshoes are meant to be lucky. But did you know that you'll only get the benefit if:
- The shoe has been worn by a horse
- It's been found and not bought
- It's positioned so the open side faces upwards (to stop the accumulated luck from falling away)

In-Car Cup Holder

TAGLINE: a hole
PREDECESSOR: the groin
LESSON: the simpler, the better

Like many of the items featured in this book, the cup holder has become so common, it hardly seems like an invention at all. It's as if the car cup holder has been around forever, like the sun or Denis Norden. Yet, it's a relatively recent invention.

You'd think someone would have thought of this immediately after man began riding on stuff. However, it wasn't until 1960 that the first auto cup holders were tested, but they didn't even catch on then. Remember, in the 1960s, seatbelts weren't standard features in cars and child car seats weren't even on the horizon. When you're busy worrying about you and your kids flying headfirst through your windscreen, I suppose your drink isn't that much of a priority. But once seatbelts and car seats became common, travelling salesmen began turning their attention to the safety of their service station coffees. Built-in cup holders began to be widely available in the 1980s, coming standard in bigger mobile caravans and high-end cars.

Then, as is often the case with well-known inventions, timing played a pivotal role in securing the cup holder's place in the international car market.

Out of nowhere came Stella Liebeck, a seventy-nine-year-old from Albuquerque, New Mexico. Stella ordered a cup of coffee from the local McDonald's drive-thru and promptly spilled it on herself. She suffered third-degree burns and sued Ronald and Company for $2.9 million. Incredibly, she won.

Realizing Stella could have just as easily sued the car company, car manufacturers the world over redoubled their efforts to provide ample cup holders in all their new vehicles from that point on. Stella, you feisty old bird, I think I speak for fluid-consuming motorists everywhere when I say, 'Thank you.' Now put that cup down, there's a dear.

Did you know?

It wasn't until 1960 that the first in-car cup holders were tested. Shortly afterwards the in-car valet service industry took a nosedive.

Pooper Scooper

TAGLINE: a scoop for picking up poop

PREDECESSOR: just walking away

LESSON: Don't be shy / give it a try / if for no other reason / than E. Coli

It's not just a matter of getting it on your shoe, or even one of common courtesy. It's a health thing. As a leading source of E. coli, your pup's poop is full of bacterial coliforms. Sure, one dog's waste might not cause global chaos and illness, but a city's worth of pets crapping in the streets every which way would create a serious health hazard. Makes you really want to clean up after your hound.

According to the website InventionConnection.com, Brooke Miller is the one who helped dog owners do just that by inventing the pooper scooper. Her original design, for which she holds the patent, is a receptacle attached to a wooden stick, with a small rake for people to use to gather the waste into the bin. It's not known for sure when Miller invented the pooper scooper, but according to InvenionConnection.com, the term was included in dictionaries in the early 1970s.

Today, nearly every urban, suburban and rural area in the civilized world has made it illegal to leave your dog mess behind in public.

> ### Did you know?
>
> Campylobacteriosis is a bacterial disease that can infect humans through contact with dog faeces. It causes mild to severe nausea, and all the good things that go along with that. The Centre for Disease Control advises pet owners to wash their hands after handling crap in order to avoid this and other bacterial diseases.

Lightning Rod

TAGLINE: a metal stick

PREDECESSOR: buildings being destroyed by lightning

LESSON: too much electricity is a bad thing

They're a patriotic bunch, Americans. Ask a tame one who his favourite founding father is and as long as he's not from a broken home or confused he'll have an answer, no doubt about it. A popular choice would be Benjamin Franklin.

He did it all. He was a publisher, inventor, scientist, politician, writer and philosopher. Although most of his inventions are ingenious and complicated – and therefore not included in this book – there is one that is simplistically brilliant: the lightning rod. Of course, at the time, inventing the lightning rod was pure genius and his experiments with electricity were unprecedented and led to a unique understanding of lightning. But come on, let's be objective here. It was a piece of metal. Of course lightning would strike it. However, without his groundbreaking lightning experimentation and vast knowledge of the

> 'An investment in knowledge pays the best interest.'
> - Benjamin Franklin

subject, we may be without this simple yet extraordinarily effective device.

Once Franklin understood that lightning is an electrical force, he thought of a very simple way to protect tall buildings from lightning strikes. All you have to do is attach a grounded metal rod to the top of each building. The lightning would be attracted to the rod instead of the structure, thereby preventing damage and fires.

Today, literally every tall structure has a Franklin lightning rod. Many of us are alive to create, invent and discover because of it. Simple yet brilliant.

Fingerprint Detection

TAGLINE: the original CSI

PREDECESSOR: a funny hat, magnifying glass and Dr Watson

LESSON: crime doesn't pay; solving crime does

People had been around for thousands of years before somebody looked down at their hand and noticed that we all have different fingerprints. Today we consider Henry Faulds, a Scottish scientist who published an article on the subject in 1880, to be the 'Father of Fingerprinting' even though he never received any real recognition for it during his day.

A few years after Faulds, Sir Edward Henry and Sir Francis Galton became interested in the technique. (Galton was Charles Darwin's cousin by the way.) Both of these men helped to create a system for 'lifting' and tracking fingerprints, but for rather different reasons. Henry was the one interested in using fingerprinting to identify criminals. Galton was more interested in using fingerprints to determine desirable characteristics in people (kind of like examining the lumps on a person's head or phrenology, which was also a popular practice during the nineteenth century). He hoped to use the technology to find the key

> **Did you know?**
>
> Like fingerprints, everyone's tongue print is unique. Just don't try licking the immigration official at the air port.

to the 'Master Race'. That's probably why modern fingerprinting is referred to as the 'Henry Method' and not the 'Crazy Racist Galton Method'. It's said that the FBI has a database of over 60 million fingerprints. Although they're beginning to be eclipsed by advances in DNA research, fingerprint files remain the most common tool for identifying people.

Assembly Line

TAGLINE: division of labour in order to increase manufacturing speed
PREDECESSOR: individual craftsmen
LESSON: grouping similar tasks makes anything go faster

Like many simple ideas, the assembly line was not 'invented' per se at one time by one person. It has been independently developed and redeveloped throughout history based on logic. However, its explosive influence in the late nineteenth century and beginning of the twentieth can be attributed to two American men – Ransom E Olds and Henry Ford – who profited greatly from this concept.

Prior to the twentieth century, manufactured products were crafted individually by hand. Individual craftsmen would use their knowledge, which was often developed through extensive apprenticeships, to make something from nothing. Some, in fact most, people, believe that this produced a higher quality product.

> 'A business that makes nothing but money is a poor business.'
> - Henry Ford

However, sometimes quantity is needed over quality.

Many think that it was Henry Ford who invented the assembly line. This isn't actually the case. What Ford did do was improve upon the concept created by Olds – founder of Oldsmobile – by implementing conveyor belts.

Incredibly, this cut the time of manufacturing a Model T Ford from a day and a half to ninety minutes. Talk about production line!

Did you know?

Henry Ford only offered his Model T in black because that was the fastest drying paint he could find.

Traffic Light

TAGLINE: traffic's guiding light

PREDECESSOR: a policeman at every intersection

LESSON: police should be catching bad guys, not standing around in intersections

Before the traffic light, a police officer would be stationed full time at junctions to direct traffic – a practice still utilized today, though only in worst-case scenarios. The first attempt to reduce the number of police needed to direct traffic was made before the motor car, when traffic consisted of pedestrians, buggies and wagons.

The first traffic light lit up in 1868 and was a revolving lantern that used red and green signals to control traffic at a London cross-roads. Yet, it still required a police officer. He had to turn the lantern so that the appropriate signal faced the right lane of traffic. Not only therefore was the officer essentially doing the same thing as he was before, but the device also had the misfortune of exploding into a fiery blaze one afternoon, injuring the policeman who was operating it and several passersby.

> 'I have not failed. I've just found 10,000 ways that won't work.'
> - Thomas Edison

Back to the drawing board.

In 1923, American Garrett Morgan was busy developing a device that could help control traffic in Cleveland, Ohio. Rather than use a lantern and policeman like the London version, Morgan decided electric and automatic was the way to go. His patent for the first successful traffic light was bought by General Electric. The company took the patent and ran, building a monopoly on traffic-light manufacturing and giving us the red-yellow-green light we know today. Well, we know red-amber-green, but same principle.

Red for stop. Green for go. And yellow/amber for speed up.

Did you know?

The average Brit spends six months of his life waiting at red lights.

Road Signs

TAGLINE: directional indications on travel routes
PREDECESSOR: asking if you can get there from here
LESSON: people generally like to know where the hell they are

The first road sign was a Roman milestone. No, I don't think you understand. They were literally Roman milestones, indicating the distance and direction to Rome. Thus the meaning of the word 'milestone' and the saying, 'All roads lead to Rome.' In the Middle Ages, these milestones evolved into multi-directional signs. Of course, sometimes they were adorned with the heads of that particular township's enemies, but nevertheless they were helpful, as established junctions along main roads like Watling Street became more common.

In 1895, an Italian Touring Club created what's believed to be the first modern road sign system. Yeah, I know – perhaps hard to believe, on the strength of today's Italian penchant for driving everywhere at maximum revs with both hands resting on the car horn. Europe's then International League of Touring Organizations advocated that a road sign system be adopted throughout the continent. However, it was never approved because it was thought to be a bit too helpful and much less entertaining for the locals – who spent most of their days amusing themselves by watching tourists wander about hopelessly lost.

These days, traffic signs are everywhere. In fact, we now have just the opposite problem: too many signs! Now we are witnessing the advent of the more commonly seen, new generation of 'intelligent' signs. This signage can actually adjust its messages and signals to adapt to road conditions and traffic flow as needed. Though those ones that flash up '30' almost always seem to get it wrong by about 10mph, I find.

Did you know?

US President Eisenhower implemented the concept of the inter-state highway system and included a federal law mandating that one in every five miles of federal highway be straight. This is so the straight sections could be used as runways during times of war or other national emergencies. This is commonly practised by the world's smaller or more easily disorientated air forces.

Windscreen Wiper

TAGLINE: device that allows you to drive in the rain without dying

PREDECESSOR: squinting and cursing and crashing

LESSON: protect your patents

While the car isn't included in this book (because it's a rather rather complicated device), just about every accessory we take for granted today has found a home in these pages, including this entry – the original windscreen wiper, which would be paired with the windscreen and replaced by the intermittent windscreen wiper.

In addition to being a lesson on accessorizing, the windscreen wiper also teaches novice inventors to check expiration dates closely (and not just on that milk that's been in the fridge for a couple of weeks). As it turns out, the original wiper patent was allowed to expire. Mary Anderson, of New York City, was granted the patent in 1903. Her original pitch to companies to begin producing the device didn't work out. But rather than keep at it, Mary put the patent in a drawer and gave up. A big mistake for Mary and any other would-be inventor out there with a dream. Rule number one: if you believe, never give up.

This allowed Fred and William Folberth to swoop in and steal Mary's squeegying thunder. The two came up with their spin on the automatic windscreen wiper in 1921 and called it the Folberths. The device quickly went from optional to standard and the Folberths started raking in the dough.

So, keep your eyes peeled for ways to build on an already existing idea, because often one great idea would not be possible without another. For example, the windscreen wiper would be rather excessive without the previous entry in this book.

Rearview Mirror

TAGLINE: a little make-up mirror in your car – don't laugh, read on

PREDECESSOR: hoping for the best when you changed lanes

LESSON: a lady is always prepared

Now that talking on a mobile phone and texting while driving is illegal, it's ironic to think about how the rearview mirror came to be.

The rearview mirror was first recommended by an early publication. *Car and Driver*? No. *Hotrod Magazine*? No. A 1906 book titled *The Woman and the Car*? Yep.

This forward-thinking, rear-looking publication was authored by Dorothy Levitt, who instructed female drivers to 'carry a little hand-mirror in a convenient place when driving' in order to 'hold the mirror aloft from time to time in order to see behind while driving in traffic'. The earliest rearview mirror that was mounted on a car rather than held daintily in one's white-gloved hand was very much big news as it was affixed to top speedster Ray Harroun's racing car for the first Indianapolis 500 in 1911. But Ray didn't point to *The Woman and the Car* as the inspiration. Instead, he said the idea came from a similar invention that was used on horse-drawn carts at the time.

The invention wasn't officially introduced until 1914, when car manufacturers began installing them. And the person who took credit for their creation wasn't Harroun or Levitt, but Elmer Berger, recognized as being the first person to note the need to install the rearview mirror during manufacturing.

So in hindsight, it seems like quite a few people can claim credit for the invention. I'm not surprised though. Often, objects in people's memories seem clearer than they appear.

Fiction is stranger than truth

Just as Dorothy Levitt wrote about the rearview mirror before it was invented, there have been many other inventions that appeared as fiction before they debuted in reality. A few of the more notable ones...

- The nuclear submarine (1955): first appeared as the *Nautilus* in Jules Verne's *Twenty Thousand Leagues Under the Sea* (1870).
- Robotics (1961): first appeared in Issac Asimov's short story *Runaround* (1942).
- The moon landing (1969): first occurred in Jules Verne's *From the Earth to the Moon* (1865).
- The mobile phone (1996): first appeared as the communications device in Gene Roddenberry's *Star Trek* (1966).

Indicator

TAGLINE: an indicator light used by conscientious drivers

PREDECESSOR: sticking your hand out the window or not at all

LESSON: remember to think and blink

This important auto feature is another case of optional going standard for safety. While first introduced in 1920 by the Protex Safety Signal Company, inventor C H Thomas claims some credit for its creation because of a *Popular Mechanics* article he wrote in 1916 that described a device consisting of a battery-run light bulb connected to a glove. Thomas's lit-up design was meant to help drivers at night, when the cars behind them couldn't see their hands sticking out their windows to signal a turn.

Buick was the first car manufacturer to feature the indicator in its automobiles. Their 1938 models came with a flashing turn signal on the rear of the car. In addition to flashing lights, some cars utilized a mini version of retractable semaphore, called trafficators, which could also be lit up for use at night or in bad visibility, but these were often prone to sticking in one position and were just too fragile.

In 1940, the blinker was also placed at the front, to help with those tricky turns across traffic.

Old school blinkers

Back before turn signals (and brake lights) became standard, drivers had to use hand signals. You probably learned (and forgot) these for your theory test. So here's a refresher:
- Straight out: right turn
- Circulating anti-clockwise: left turn
- Raising and lowering a straight arm repeatedly: slowing down to stop
- Elbow bent, middle finger raised: get off my arse

Barbed Wire

TAGLINE: a really pointy fence
PREDECESSOR: fences that didn't hurt
LESSON: better fences make better neighbours

Have you ever urinated outdoors? No? For the sake of brevity, let's dispense with the formality of pretending you're not lying. As someone who's peed outdoors, you've probably been in a scenario where you've just 'got to go' so you creep into the undergrowth only to be stopped by thorny bushes and are forced to go elsewhere. Essentially, that is how barbed wire works, only without the piss.

People are often surprised when told that barbed wire is considered one of the most significant inventions of the past 200 years. After all, it's synonymous with mud-churned battlefields, concentration camps and painful attempts to exercise your right to roam across farmland. But barbed wire is held in this high regard, at least in the US, for one reason: the cow.

> 'Careful as a naked man climbin' a barbed wire fence.'
> - Cowboy proverb

See, back in the days of the American Wild West, livestock grazed freely. Before the introduction of the 'thorny fence' (as barbed wire is also known), wild and domesticated animals simply penetrated existing fence systems and had their way with crops. Think about it. If one little bunny can gnaw his way through your carefully cultivated lettuce patch, imagine the damage that could be done by 10,000 head of 1,500-pound cattle! It was an American called Lucien B Smith who helped rein in the livestock. He received the first patent for barbed wire in 1867. Joseph F Glidden improved on the concept and was issued a patent for his modified version in 1874.

The widespread use of this highly effective fencing method changed life in the west almost as dramatically as line dancing and the gigantic belt buckle. Without this extraordinarily simple invention, US agriculture would have been severely stunted, making western migration and the settlement of the majority of the United States impossible. Though maybe there'd have been more cowboys employed to keep the burger-meat in check.

Lever

TAGLINE: a prying and lifting device
PREDECESSOR: hernias
LESSON: sometimes inventions are so simple they just occur naturally, so keep your eyes peeled

What do you say when someone has an advantage over a particular situation? Well, you might just say they have leverage. Why? Because the word's derived from lever, which provides an enormous advantage in power over anything upon which it is used.

Got it? Good. Let's think about the lever's use by taking a stroll down memory lane to a more innocent time when you spent your younger days on the playground. You see another kid sitting on one end of the seesaw. You run over to the opposite side, looking to climb on. That other kid is kind of big though. You're worried you won't be able to move her an inch off the ground. But miraculously when you climb on, she rises into the sky and then she comes back down and you rise up. The power of the lever! And then she jumps off while you're still lifted and you crash to the ground for the tenth time that day even though she promised she wouldn't do it again so you spend the next twenty-five years of your life harbouring mistrust issues and resentment and... ahem... sorry. Back to the power of the lever!

The seesaw is a simple example of how this simple machine works. The lever is one of the most useful basic tools. While the lever itself isn't an actual 'invention,' the concept has been utilized in a number of important ones. It's helped do everything. The ancient Egyptians used levers to raise tremendous stones, sometimes weighing more than 200 tons. As for a modern use – just think of how much more weight you can move in a wheelbarrow.

Why? Because of the lever! And many other basic tools we use everyday are levers, including scissors, pliers, hammer claws, nutcrackers and tongs. But my favourite use will always be the seesaw! (Oh and I'll get you back, Tracy Wilkinson – it may have been in the fourth year juniors, but I remember… oh I remember…)

> 'Give me a place to stand, and I shall move the earth.'
> - Archimedes

Standardized Time

TAGLINE: time so everybody knows what time it is at the same time

PREDECESSOR: I'll tell you in exactly one minute

LESSON: use your head and your watch

Standard time was implemented first in Britain and was called 'railway time'. Why? Because in 1840 the Great Western Railway decided to synchronize a number of local times along its route and work its timetables to a new unified clock set to Greenwich Mean Time (GMT). For instance, Bristol had been ten minutes behind London. Quickly other railway lines adopted the plan (good job too – imagine the carnage otherwise), and as railways and industrialization spread within the British empire and in countries like France and the US, so the agreed times governing daily rhythms were established. The ingenious plan in place before that? Locality and winging it.

See, before time became standardized, it was approximate, by locality, believe it or not. Each town maintained its

> 'There was even railway time observed in clocks, as if the sun itself had given in.'
> - Charles Dickens, *Dombey and Son*

own time by means of a local, well-known clock. For example, sometimes it was a public clock maintained by the town (think Big Ben). Sometimes it was simply set by the clocks in an established clockmakers or jewellery shop window.

> **Did you know?**
>
> A 'jiffy' is an actual unit of measurement. It was invented in early 1900s by American chemist Gilbert Newton Lewis. In fact, the jiffy is exactly 1/100th of a second, though it has other measurements depending on the field of study to which it is being applied.

Okay, it was roughly the same in a small country like Britain, and new technologies such as the telegraph helped matters, but time zones were difficult to account for and often ignored completely. Live in a bigger country and your problems are magnified, e.g. in the United States, which didn't mandate standardized time in different longitudes until 1918.

As populations grew and technology advanced, the use of standard time gradually became more common and necessary – which, in turn made communication and travel more efficient. It's an interesting argument whether industrialization created standardized time, or whether standardized time paved the way for industrialization.

A MATTER OF LIFE AND DEATH

Guillotine

TAGLINE: head severing device

PREDECESSOR: hanging, stoning, bludgeoning, drowning, pummelling, hacking

LESSON: if you want 'em dead, find a tidy way to behead

Gruesome? Of course. But the cold hard fact is: humans have been trying to find simple, efficient ways to execute one another since before recorded time.

And a lot of recorded times it didn't go all that well. Hanging

Axe for a Better Way to Go?

In 1541, it took three strokes of an axe to sever the head of Margaret, Countess of Salisbury. Trees have been brought down quicker than that.

> **Did you know?**
>
> While the guillotine is probably best known for its use during France's 'reign of terror' (when it was used on King Louis XVI and Marie Antoinette), the machine was also employed during the Second World War. The Nazis used the guillotine on 16,500 people in Germany and Austria between 1933 and 1945.

was the most popular form of execution and definitely no picnic. To note just one of many examples, witnesses received quite a surprise at the 1906 execution of one William Williamson. When the trap door swung open, William fell all the way to the ground and landed on his feet. (Seems his executioner was rat-arsed.) Can you imagine that? You're waiting to have your neck snapped and you land on your feet. Bonus! That's a Get Out of Jail Free card, right? Wrong. Three deputies, standing on the scaffold, seized the rope and forcibly pulled William off the floor for fourteen and a half minutes until the coroner pronounced him dead from strangulation. Nice attitude, guys.

Countless instances of human error in legally snuffing people out led Tobias Schmidt, a German engineer, to let gravity and a blade do the work when he invented the guillotine in the eighteenth century.

Heimlich Manoeuvre

TAGLINE: squeezing bits of food out of embarrassed diners

PREDECESSOR: inadvertently jamming the foreign object further down people's throats by pounding on their backs

LESSON: never breathe with your mouth full. Or talk. Or eat.

Ronald Reagan; Cher; Halle Berry. What do these people have in common? Dreadful films in their back catalogues? Well, yes, but not only that. They've all allegedly been saved from choking by one Henry Heimlich, MD, who introduced the procedure in 1974.

While the footage would make fascinating YouTube fodder, these famous folks aren't the only people who've inadvertently sucked down hunks of food into their larynxes. Plenty of average people have managed to send their suppers down the wrong pipe, only to be saved by the Heimlich manoeuvre.

It's estimated that the Heimlich has saved more than 50,000 people. Why? Because people talk with their mouths full, that's why! Listen to your mothers! So you know that the Heimlich works, but what about how it works? Well it's as simple as flobbing a chewed-up bit of paper through a straw at someone.

The soggy projectile goes into the straw at one end, then it's shot across the room by blowing through the other. Just like this, the food goes down the choker's windpipe and then is shot out when the squeezer grabs the choker from behind and applies pressure to the diaphragm with an abominable thrust. Simple as that. Though normally it's not paper that comes back up.

Did you know?

You can perform the Heimlich manoeuvre on yourself. Ball your fist and place it just above your belly button and then thrust it into your abdomen using your other hand; or you can set yourself against the edge of a table, desk or back of a chair (again, just above your belly button) and then thrust yourself into it. You'll look some kind of deviant either way, but who cares if you happen to be choking at the time?

Plaster

TAGLINE: a little piece of tape with some gauze stuck to it
PREDECESSOR: tiny wound overkill
LESSON: stop blood loss

We all know somebody who's clumsy – banging into stuff, falling down a lot – the kind of person who scares the hell out of you when he handles a knife or, God forbid, does that leaning-back-in-a-chair thing.

And, we all know a Boy Scout or two.

Well, in 1917, that's how the Elastoplast, or sticking plaster to you and me, came to be. A cotton buyer named Earle Dickson married a clumsy bird named Josephine Frances Knight who continually wounded herself. Earle would end up patching Josephine up with tape and some gauze from the company for which he procured cotton, Johnson & Johnson. Problem was, the stuff was just too big for her cuts and scratches. Poor Earle felt like he was using a cannon ball to kill a fly.

So Earle made a batch of little bandages by affixing pre-cut squares of Johnson & Johnson sterile gauze to some little bits of surgical tape. He found some fabric called Conaline, which he used to cover the sticky parts of the tape so the bandages wouldn't start sticking before they were supposed to.

The little devils worked so well that he presented the idea to his boss, James Johnson, who ordered production to begin in 1920. At first, sales were slow. Then, somebody had the idea to distribute free samples to Boy Scout troops, which, as you know, are just teeming with eager boys running around in sharp, pointy forests wearing shorts.

Brilliant! The little invention took off and, to date, more than one hundred billion Band-Aids have been sold. Well done Earle, well done. The British version, Elastoplast, was created in 1928 by Smith and Nephew, a company founded way back in 1856 which today prides itself on being the world's second biggest wound management expert. That's a bloody big wound – let's hope they've kept a few of the jumbo-sized plasters back in case they come across the world's biggest.

Seatbelt

TAGLINE: helps motorists stop going 70 mph when their cars do

PREDECESSOR: an untimely introduction to your windscreen

LESSON: use your head to make an impression on something other than the dashboard

Seatbelts are one hell of a lifesaver, aren't they? So logical and simple you'd think they would've come standard with the wheel. But, no. Why? Because no one capitalized on the opportunity. In fact, it took until the 1950s for seatbelts to start showing up in cars. That's over fifty years after the first automobile was invented in 1870.

Even after manufacturers were required to install seatbelts in the 1960s, people didn't use them. I remember back when I was a kid in the '70s and '80s, my parents would drive along the motorway at 75 miles per hour while my friends and I sprawled around freely in the back seat, unbelted, smoking cigars, drinking beer and gambling on dog races. Man, those were the days.

The invention of the over-the-lap seatbelt is attributed to English engineer George Cayley; he came up with the concept in the late 1800s. However, the patent for the first car seatbelt belongs to Edward J Claghorn, an American. His patent was issued in

> 'A lot of the people who keep a gun at home for safety are the same ones who refuse to wear a seatbelt.'
> - George Carlin

1885. And it showed up first in American models by Nash and Ford, but only as optional add-ons. Swedish manufacturer Saab was the first to place them in cars as standard equipment in 1958.

It took a few years for seatbelts to become standard in vehicles and it took even longer for them to be used regularly. Wearing a seatbelt didn't start to become mandatory until 1983, when Parliament passed legislation making it illegal to ride in the front of a car without wearing your seatbelt. The back seat passengers could still take the risk legally up until 1991. Since then, seatbelts have saved millions of lives, are standard in every car and their use is mandatory nearly everywhere – though in the United States seatbelt law varies from state to state, such that some territories demand buckling up in the front but not the back and, in the case of New Hampshire, no legal requirement exists at all.

Theory of Evolution

TAGLINE: only the strong survive

PREDECESSOR: God

LESSON: sometimes the greatest ideas in the world are right before our eyes the whole time

Everybody's all, 'Oh! The theory of evolution! How groundbreaking and clever!' Erm... have you ever looked at a monkey and noticed how much they look like us? It's just that simple really. (Well, obviously it isn't as simple as that, but you know what I mean.)

Not to say that I could've thought it up. But come on. This elegant theory is a perfect example of how such an amazingly obvious idea can go unnoticed for centuries until one person comes along and gets famous from it.

But did you know that Charles Darwin wasn't the first person to bring it up? Revolutionary evolutionary ideas have been around since the sixth century BC. Back then, Greek philosopher Anaximander was examining and expanding on theories of evolution. And he wasn't the only one. Greek philosophers Empedocles and Lucretius, Arab biologist Al-Jahiz and Persian philosopher Ibn Miskawayh all threw their hats into the evolutionary ring.

> 'Owing to this struggle for life, any variation, however slight and from whatever cause proceeding, if it be in any degree profitable to an individual of any species, in its infinitely complex relationship to other organic beings and to external nature, will tend to the preservation of that individual, and will generally be inherited by its offspring.'
> - Charles Darwin, *On the Origin of Species*

And the list doesn't even end there (though for the purposes of this book, it does) and it doesn't even begin to include all the indigenous people of various lands who have woven this theory into their cultural and religious belief systems for thousands of years undocumented.

However, as far as western civilization is concerned, it was Charles Darwin who popularized the idea that we evolved from apes. His *On the Origin of Species* brought the idea of natural selection into the collective conversation. The 1859 publication still stands as the bible of evolutionary theory to this day.

Electric Chair

TAGLINE: the last uncomfortable seat you'll sit on
PREDECESSOR: chairs that didn't kill you
LESSON: look for efficient 'painless' ways to kill one another

Unpleasant as this may be, it's a very simple and good idea, at least to the judiciary in those states of the US which still use it. As is the case with the guillotine, people have been trying to find simple, efficient ways to execute one another for thousands of years. And it's not as easy as it seems. For example: take this scene from a Texas jailhouse on 24 June, 1987 – in an effort to administer a lethal injection to Elliott Johnson, it took thirty-five minutes to insert a needle into his vein.

Hanging (the most popular form of execution) was no picnic either. A good deal of experimentation had to be done to determine the proper drop-to-weight ratio. Too short a drop, the neck isn't snapped and the victim dies a horribly slow strangulation. Too long a drop and the victim's head pops off like a cork.

So, Harold Brown had the idea that electricity would be the new and improved, humane way to get the job done. To prove the efficiency of his electric chair in order to secure a patent, Brown started demonstrating the machine's killing ability by using it on cats and dogs. Even with his trial runs on Fluffy and Rover, the New York Commission still wasn't sold on the idea. So Brown fried himself up some steak. He brought a cow before the panel of patent officials and flipped the switch. But Brown's animal kingdom killing spree didn't stop there. He also frazzled a horse, an elephant and an orang-utan – which ended up catching on fire. (Blimey, he's lucky PETA wasn't around back then!) Finally the commission was convinced of the invention's efficiency and the electric chair became an official means of execution on 4 June, 1888. Reliable statistics are surprisingly thin on the ground, but since 1890 over 4,500 people have been executed by electric chair in the US.

You can order anything as long as it's fried

The US is the only nation which officially still 'chairs' convicted criminals, though many states reject the method as inhumane and have banned it in favour of lethal injection or no death penalty at all. But the good folk of Alabama, Florida, South Carolina and Virginia still offer 'Old Smokey' on the menu. Notoriously variable in its efficiency, the electric chair has even been applied twice to some prisoners whom the authorities failed to dispatch properly first time round.

Sterile Medical Procedures

TAGLINE: washing one's hands before shoving them into someone else's body

PREDECESSOR: killing people

LESSON: scrub up, pup

It's amazing what we take for granted these days. We should remember that it wasn't so long ago when performing medical procedures and operations in a sterile medical environment was barely a consideration. In the early nineteenth century, surgeons operated in the clothes they wore to work that day, not even bothering to necessarily put on a lab jacket or apron, let alone a mask. In many cases, butchers were more conscious of being sterile. And the surgical dressings surgeons used on their patient's wounds were made of uncleaned cotton swept from textile mill floors.

Shockingly, post-surgical mortality in the 1800s was as high as 90 per cent due, mostly, to these unsterile conditions. What makes this statistic even more shocking is that all you have to do in order to sterilize a medical instrument is drop it into a pot of boiling water. So why didn't they do that? Why indeed. It was simply because someone, like you, just didn't think of it.

English surgeon Joseph Lister was the first to properly link operating room infection to germs while practising at the Glasgow Royal Infirmary. Building on some of the principles laid out by Louis Pasteur, he experimented with the eradication of bacteria through chemical and heat treatment; his primary discovery was the sterilizing effect of carbolic acid. But he also spotted what seems to us to be a self-evident truth; that washing one's grubby mitts cuts down on the amount of dirt going round.

He noted that the infant mortality rates of children born with the help of a midwife were significantly lower than those born under the supervision of a surgeon. The simple reason for this disparity was that midwives washed their hands more often, whereas surgeons might turn up to assist a birth having just dug out an infected wound from a tramp in the ward next door. Thus Lister is hailed as the 'father of modern antisepsis'. What better way to commemorate his achievements than with a mint-flavoured mouthwash and a food-poisoning disease? Step forward Listerine and listeria.

Soon after Lister's discoveries, companies began to develop antiseptic wound dressings. This breakthrough was the beginning of an age that sought to improve patient care through the use of sterilization and disinfection technologies. Since then things have improved dramatically in this area and pre-surgical sterilization has became a standard during all medical procedures, saving countless lives.

Condom

TAGLINE: a super-elastic, ultra-fantastic prophylactic
PREDECESSOR: disease and unwanted children
LESSON: sex sells

Let's take a good long hard look at condoms, shall we? Believe it or not, these things have been around forever.

People who study this sort of thing have found that the use of condoms began around 1000 BC. However, unlike today's latex or rubber varieties, early condoms were made of linen sheaths, leather, oiled silk paper, or from a very thin hollowed-out horn.

As the story goes, the condom got its name because the popular sheep entrails variety was first used by farmers in the French town of Condom. While you might find using a sheep's innards as a contraceptive revolting, at least it was only intended for a single use. When the less-disgusting rubber condoms came on the scene in 1855, due to their cost, people were advised to wash and reuse them until they fell apart. Now, what's really less disgusting? I'd say it's a toss-up.

The latex condom didn't pop up until 1912. The use of latex made the product both affordable and disposable. Goodbye sheep guts and wash-and-wear rubbers. The modern condom was born. But it took until the 1980s and the spread of HIV for people to get over the moral concerns that hampered the invention's acceptance and use. Thankfully, today it's concerning if you don't use a condom. And production rates show that people are using them. Though the Pope still has something to say about that.

The United Nations Population Fund reports that over ten billion condoms were used in 2005. And nowadays, you can find them in a variety of colours, textures and even flavours.

Snack time

Condom flavours include banana, chocolate, strawberry, raspberry, melon, orange, grape, mint, cola, vodka, bacon and egg, doner kebab and even good old latex for unadventurous types.

Mouse Trap

TAGLINE: cute, little, rodent-crushing device
PREDECESSOR: more mice with fewer spinal injuries
LESSON: your idea already exists? Build a better one

In 1897, Leeds-based inventor James Henry Atkinson designed what we've come to know as the common mouse trap. But that's not to say that people didn't try to trap mice before 1897. There were all kinds of trap contraptions that existed before (and after) Atkinson's – they just weren't as good at catching mice. And while there are literally thousands of varieties (and one great but fiddly board game), his version remains the most popular.

Atkinson's design is the one that sits on a little plank of wood and has a powerful spring-loaded, catapult-type mechanism. A piece of food is placed on the trap and when the mouse goes to get it, he triggers the release of a metal bar, which hopefully (but not always) kills the mouse by shattering its skull, snapping its spinal cord, or breaking its ribs and crushing its lungs. The poor little bugger won't know what's hit him! And Atkinson called it the Little Nipper. Ahhh! It sounds so cute. Don't be fooled though – it's a killing machine.

Atkinson sold the patent for his mouse trap to William Procter in 1913 for £1,000. The Procter Bros company has been manufacturing – and making money from – his Little Nipper ever since. Atkinson's mouse exterminator is so successful that Procter has erected a 150-exhibit mouse trap museum in its factory.

All told, the Atkinson mouse trap is the winner among many, as over 4,400 patents have been issued for various types of mouse traps, including electrocuting traps, glue traps, bucket traps, humane traps and so on. And only twenty have made money – which leads us to Ralph Waldo Emerson's classic remark... 'Build a better mouse trap and the world will beat a path to your door.' I can't think of a better way to sum up the spirit of this book.

Thanks Ralph!

Atkinson's Little Nipper slams shut in 1/38,000th of a second. Just ask any mouse who looks like he can dodge in 1/38,001st of a second.

Resources

American Inventors www.american-inventor.com

Brainy Quote www.brainyquote.com

British inventions www.inventors.about.com/library/inventors/bl_british_inventors.htm

Discovery Channel www.dsc.discovery.com

EduMedia www.edumedia-sciences.com

eHow www.ehow.co.uk

Famous Black Inventors www.black-inventor.com

Famous Women Inventors www.women-inventors.com

Innovation UK www.innovation.uk.org

Invention Connection www.inventionconnection.com

Inventions – Inventors and Inventions – Famous Inventors www.topinventionsinfo.com

Inventorprise Inc. www.inventorprise.com

Inventors Digest www.inventorsdigest.com

Science www.sciencemag.org

ThinkQuest www.library.thinkquest.org

United Inventors Association (UIA) www.uiausa.com

United States Patent and Trademark Office (USPTO) www.uspto.gov

Wikipedia www.wikipedia.org

Wise Geek www.wisegeek.com

Appendix

Resources for the At-Home Inventor

To Inspire: Check out these sites to see if your idea already exists, or just for inspiration.

Advertising Age www.adage.com
Alexa www.alexa.com
Good Ideas Salons www.goodideassalons.com
Google Trends Labs www.google.com/trends
Inventors Digest www.inventorsdigest.com
Kuukan.com www.kuukan.com
Popular Science www.popsci.com
Springwise www.springwise.com
Squidoo www.squidoo.com
Trendwatch www.trendwatching.com
Why Not www.whynot.net

To Educate: Have a great idea, but don't know the next step? Check out these websites to get on the right path...

The Academy of Applied Science www.aas-world.org
Ask the Inventors! www.asktheinventors.com
British Inventors' Society www.thebis.org
Delphion www.delphion.com
Intellectual Property Office (formerly UK Patent Office) www.ipo.gov.uk
International Federation of Inventors' Associations www.invention-ifia.ch
Inventique www.wrti.org.uk
Inventors HQ www.inventorshq.com
InventorEd www.inventored.org
Patent Café www.patentcafe.com
United Inventors Association www.uiausa.org
United States Patent and Trademark Office www.uspto.gov

About the Author

Anthony Rubino Jr was born in New Jersey to a first-generation Italian-American family. Needless to say he developed a sense of humour at an early age... and then felt guilty about it. Channelling that early confusion, he now combines humour, art and pop culture to create drivel of the highest quality.

Never a stickler for maths, Tony wrote five books for his *Life Lesson* 'trilogy': *Life Lessons from Your Dog*, *Life Lessons from Your Cat*, *Life Lessons from Elvis*, *Life Lessons from the Bradys* and *Life Lessons from Melrose Place*. Before that he displayed his steely work ethic by penning *1001 Reasons to Procrastinate*. And his fear of the discomfort of eternal damnation is reflected in his *The Get Into Heaven Deck: Or Your Money Back*. Along the way Tony has contributed his articles and cartoons to publications such as *MAD Magazine*, *Cracked*, *National Lampoon*, the *Chicago Tribune* and *Opium Magazine*.

He is currently writing the daily cartoon strip 'Daddy's Home' which appears in more than 250 newspapers and websites (www.daddyshomepage.com). When not working on his writing and art in New York City, he spends his time not working on his writing and art in New York City. Visit www.rubinocreative.com for more information and big, big fun.

Index

3M Company 56–7, 78–9

Aiken, Howard 71
Al-Jahiz 144
Alliance Rubber Co 73
Anaximander 144
ancient Egypt 84–5, 100, 133
ancient Greece 46, 48, 84–5, 144
Anderson, Mary 124–5
Archimedes 133
Asimov, Isaac 127
assembly lines 118–19
Atkinson, James Henry 152–3
Ayscough, James 106–7
Aztecs 8, 24

B F Goodrich Company 103
Ball, Richard 38
barbed wire 130–1
Barbie doll 50–1
bell-ringing 54–5
Berger, Elmer 127
Berry, Halle 138
Big Ben 55
bikinis 96–7
Bild Lilli doll 50–1
bottled water 28–9
Boy Scouts 140, 141
bras 94–5
bread, sliced 16–17
Brondel, J F 64
Brown, Harold 146–7

Brown, Oscar 109
Bryant and May 87
Burden, Henry 109

Cadbury 25
campylobacteriosis 113
carbolic acid 149
Carlin, George 143
carrier bags 82–3
cattle 130–1
Cayley, George 142
Chapelle, Francis 29
Cher 138
Chicago Roller Skate Company 53
chimpanzees 10
China 46, 48, 88–9, 106
chocolate bars 24–5
Claghorn, Edward J 142–3
coat hangers, wire 70–1
condoms 150–1
Copernicus, Nicholas 15
Cortez, Hernando 24
Cowell, Simon 101
Crapper, Sir Thomas 65
Crocs 92–3

Darwin, Charles 144–5
Davison, Alvin 76–7
de Mestral, George 104–5
defence 54–5
Diamond Match Company 87
Dickens, Charles 134
Dickson, Earle 140–1
drawing pins 80–1
Drew, Dick 79
drums 42–3
duct tape 58–9
Duncan, D F 35
Duracell 91
Durand, Peter 60, 61

earmuffs 98–9
ecstasy 38
Eddison, Alan 90
Edison, Thomas 69, 120
Edward III 47
Einstein, Albert 14, 15
Eisenhower, Dwight D 123
electric chair 146–7
Elizabeth I 54–5
Emerson, Ralph Waldo 153
Empedocles 144
Epperson, Frank 18–19
evolutionary theory 144–5
executions 136–7, 146–7

Faber-Castell 74
Faulds, Henry 116
Fay, Samuel B 68–9
FBI 117
fighter pilots 106, 107
fingerprinting 116–17
fire 5, 6–7, 86, 87

INDEX 157

Fish, Michael 101
fishing rods 48–9
Flambeau Plastic Company 35
Folberth, Fred 125
Folberth, William 125
football 46–7
Ford, Henry 118–19
Forster, Charles 66–7
Foster, Sam 107
Franklin, Benjamin 114–15
Franscioni, Warren 40
French fries 22–3
Frisbees 40–1
Fry, Art 56–7

Galileo Galilei 15
Galton, Sir Francis 116–17
General Electric 121
Glidden, Joseph F 131
global warming 82
Goldman, Sylvan 32–3
golf 5, 44–5
Gore, Al 82
graphite 74–5
gravity 14–15
Great Mingun Bell 55
Greenwich Mean Time 134
Greenwood, Chester 98–9
guillotine 136–7

haggis 21
hand signals 129
Handler, Barbara 50–1
Handler, Ruth 50–1
hanging 136–7, 146
Harrington, Sir John 64
Harroun, Ray 126–7
Heim, Jacques 96, 97

Heimlich manoeuvre 138–9
Henry, Sir Edward 116–17
Herbert's Speciality Meats 21
highways, straight 123
hippies 39
HIV 150–1
horseshoes 108–9
Hugo, Victor 76
Hula Hoops 36–7, 41

Ibn Miskawayh 144
ice lollies 18–19
in-car cup holder 110–11
Inca people 8
indicators 128–9

James I of Scotland 47
James II of Scotland 45
Jefferson, Thomas 22–3
jiffy 135
Johnson, Elliott 146
Johnson, James 141
Johnson and Johnson Permacel Division 59
Johnson & Johnson 140
Jones, Samuel 86
Joseph Fry & Son 24–5
Judson, Whitcomb 102
Julien, Honoré 23

Kaggwa, Stephen 79
Ken doll 50
Kim Jong-il 45
Kirsten, Johann 81
Knerr, Richard 41
Knight, Josephine Frances 140

Knight, Margaret 83
knives 12–13

lethal injection 146, 147
levers 132–3
Levitt, Dorothy 126–7
Lewis, Gilbert Newton 135
Liberty Bell 55
Liebeck, Stella 111
lightning rod 114–15
Linstedt, Otto 81
Lister, Joseph 148–9
Lucretius 144
Lundstrom, Johan Edvard 86–7
Lyman, William 61

Macaronis 100–1
Mallett, Timmy 97
Margaret, Countess of Salisbury 136
masking tape 78–9
matches 86–7
Mattel 50–1
Mayan people 8
McDonald's 111
McKnight, William 79
medical procedures, sterile 148–9
Mellin, Arthur 'Spud' 41
membranophones 42–3
milk crates, plastic 26–7
Miller, Brooke 112
mobile phones 127
Montezuma 24
moon landings 127
Moore, Edwin 80
Morgan, Garrett 121
Morrison, Frederick 40–1

mouse traps 152–3
Museum of Modern Art 57

Nazis 137
Neptune 14
Nestlé 25
Newton, Sir Isaac 14, 15
nuclear submarines 127

Olds, Ransom E 118–19
orang-utans 10

paper bags 83
paper cups 76–7
paper towels 62–3
paperclips 68–9
Parkhouse, Albert 70–1
Pasteur, Louis 149
pencils 74–5
Perrier 29
Perry, Stephen 72
Peter, Daniel 25
Philippines 35
pillows 84–5
plasters 140–1
Pluto Platter 41
police 120
pooper scoopers 112–13
Post-it Notes 56–7
Premiership football 47
Proctor, William 153
Protex Safety Signal Company 128
Ptolemy, Claudius 14

railways 134
Rainbow (TV show) 103
Reagan, Ronald 138
Reard, Louis 96–7
rearview mirrors 126–7
recycling 90–1

Richard, Bill 53
road signs 122–3
robots 127
Roddenberry, Gene 127
Rohwedder, Otto Frederick 16–17
Romans 28, 46, 67, 84–5, 88–9, 108, 122
Royal Navy 60, 61
rubber bands 72–3

Schmidt, Tobias 137
Scialabba, George 97
Scotland 44–5
Scott Paper Company 62, 63, 89
seatbelts 142–3
Second Summer of Love 38
Second World War 59, 90, 137
shopping trolleys 32–3
Silver, Spencer 56–7
skateboards 52–3
smiley face icon 38–9
Smith, Lucien B 131
Snickers 24
Spain, Bernard 29
Spain, Murray 29
spears 10–11
Spencer, William 72–3
sterile medical procedures 148–9
Stone Age 7, 8
Stone, Marvin 30–1
straws 30–1
submarines, nuclear 127
Sundback, Gideon 102
sunglasses 106–7
Swan Vesta 87
swords 13

Taft, William 87
Tell, William 15
Thomas, C H 128
ties 100–1
Timberlake Wire and Novelty Company 70–1
time, standardized 134–5
tin openers 60–1
toilet bowls 64–5
toilet paper 88–9
tongue prints 117
toothpicks 66–7
traffic lights 120–1
Tsar Bell 55
Tucek, Marie 95
turducken 20–1

UEFA 47
Uranus 14

Val Surf Shop 53
Velcro 104–5
Verne, Jules 127
vinyl 26

Walker, John 86
Walker, Col. Lewis 102
Warner, Ezra 60–1
water, bottled 28–9
Wham-O 36–7, 41
wheel 5, 8–9
Williamson, William 137
windscreen wipers 124–5
wire coat hangers 70–1
Wolle, Francis 83
World Peace Bell 55

yo-yos 34–5

Zippy 103
zips 102–3

Loved this book?

Tell us what you think and you could win another fantastic book from David & Charles in our monthly prize draw.

www.lovethisbook.co.uk

Forbidden Knowledge: 101 Things Not Everyone Should Know How to Do
Michael Powell
978-1-5986-9525-0
Want to beat a lie detector test, count cards at a casino or even start a riot? If so, this book really is your ultimate guide to living dangerously and doing all the things you shouldn't be.

747 Things to Do on a Plane
Justin Cord Hayes
978-1-5986-9541-0
Make the most of your mile-high experience and be bored no more! This travel companion features everything from counting passengers to in-chair aerobics and even romance at 30,000 feet.

Weedopedia
Will B. High
978-1-4405-0645-1
Featuring over 800 entries from how to make a bong out of a coconut to the real way to get stoned at festivals, this is every stoner's one-stop shop to everything weed. After all, there's more to marijuana than smoking it!

Whoogles
Kendall Almerico
978-1-4405-1086-1
If you've ever thought your google enquiries were strange, think again. This bizarre collection of 100% real google searches reveals the strange, ill informed and sometimes sick mind of everyday people.